This guide is dedicated to the memory of Andy Sasiadek, biology teacher and OCR examiner.

Philip Allan, an imprint of Hodder Education, an Hachette UK company, Market Place, Deddington, Oxfordshire OX15 0SE

Orders
Bookpoint Ltd, 130 Milton Park, Abingdon, Oxfordshire OX14 4SB
tel: 01235 827827
fax: 01235 400401
e-mail: education@bookpoint.co.uk
Lines are open 9.00 a.m.–5.00 p.m., Monday to Saturday, with a 24-hour message answering service.
You can also order through the Philip Allan Updates website: www.philipallan.co.uk

ISBN 978-1-4441-6257-8

First printed 2012
Impression number 5 4 3 2 1
Year 2017 2016 2015 2014 2013 2012

Cover photo: Fotolia

Printed in Dubai

Hachette UK's policy is to use papers that are natural, renewable and recyclable products and made from wood grown in sustainable forests. The logging and manufacturing processes are expected to conform to the environmental regulations of the country of origin.

P02022

Contents

Getting the most from this book ... 4
About this book ... 5

Content Guidance

Cellular control and variation ... 7
Meiosis and variation ... 15
Selection ... 28
Biotechnology and gene technologies ... 35
Cloning ... 35
Biotechnology ... 38
Genomes and gene technologies ... 41
Ecosystems ... 51
Responding to the environment ... 60
Plant responses ... 60
Animal responses ... 62
Animal behaviour ... 69

Questions & Answers

Q1 Gene control ... 74
Q2 Gene interaction ... 78
Q3 Genomes, gene sequences and selection ... 83
Q4 Cloning ... 89
Q5 Ecosystems: fieldwork ... 91
Q6 Plant responses ... 95
Q7 Animal responses ... 100

Knowledge check answers ... 105
Index ... 107

Getting the most from this book

Examiner tips

Advice from the examiner on key points in the text to help you learn and recall unit content, avoid pitfalls, and polish your exam technique in order to boost your grade.

Knowledge check

Rapid-fire questions throughout the Content Guidance section to check your understanding.

Knowledge check answers

1 Turn to the back of the book for the Knowledge check answers.

Summary

Summaries

- Each core topic is rounded off by a bullet-list summary for quick-check reference of what you need to know.

Questions & Answers

Exam-style questions

Examiner comments on the questions

Tips on what you need to do to gain full marks, indicated by the icon e.

Sample student answers

Practise the questions, then look at the student answers that follow each set of questions.

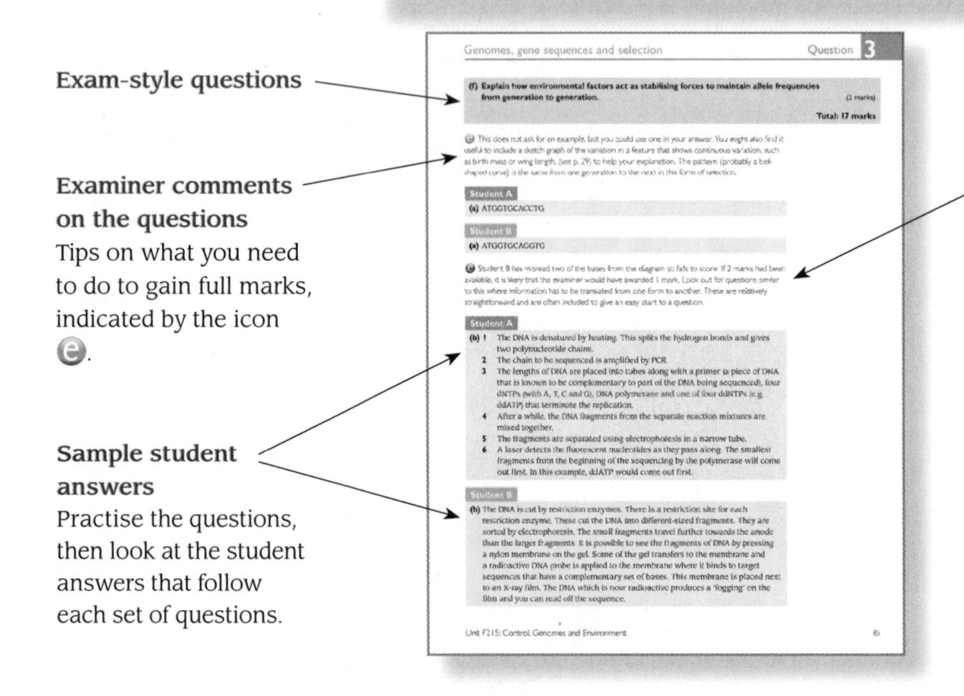

Genomes, gene sequences and selection — Question 3

(f) Explain how environmental factors act as stabilising forces to maintain allele frequencies from generation to generation. (3 marks)

Total: 17 marks

e This does not ask for an example, but you could use one in your answer. You might also find it useful to include a sketch graph of the variation in a feature that shows continuous variation, such as birth mass or wing length, (see p. 29) to help your explanation. The pattern (probably a bell-shaped curve) is the same from one generation to the next in this form of selection.

Student A

(a) ATGGTGCACCTG

Student B

(a) ATGGTGCAGGTG

e Student B has misread two of the bases from the diagram so fails to score. If 2 marks had been available, it is likely that the examiner would have awarded 1 mark. Look out for questions similar to this where information has to be translated from one form to another. These are relatively straightforward and are often included to give an easy start to a question.

Student A

(b) 1 The DNA is denatured by heating. This splits the hydrogen bonds and gives two polynucleotide chains.
2 The chain to be sequenced is amplified by PCR.
3 The lengths of DNA are placed into tubes along with a primer (a piece of DNA that is known to be complementary to part of the DNA being sequenced), four dNTPs (with A, T, C and G), DNA polymerase and one of four ddNTPs (e.g. ddATP) that terminate the replication.
4 After a while, the DNA fragments from the separate reaction mixtures are mixed together.
5 The fragments are separated using electrophoresis in a narrow tube.
6 A laser detects the fluorescent nucleotides as they pass along. The smallest fragments from the beginning of the sequencing by the polymerase will come out first. In this example, ddATP would come out first.

Student B

(b) The DNA is cut by restriction enzymes. There is a restriction site for each restriction enzyme. These cut the DNA into different-sized fragments. They are sorted by electrophoresis. The small fragments travel further towards the anode than the larger fragments. It is possible to see the fragments of DNA by pressing a nylon membrane on the gel. Some of the gel transfers to the membrane and a radioactive DNA probe is applied to the membrane where it binds to target sequences that have a complementary set of bases. This membrane is placed next to an X-ray film. The DNA which is now radioactive produces a 'fogging' on the film and you can read off the sequence.

Unit F215: Control, Genomes and Environment — 85

Examiner commentary on sample student answers

Find out how many marks each answer would be awarded in the exam and then read the examiner comments (preceded by the icon e) following each student answer.

About this book

This unit guide is the second of two that cover the OCR A2 specification in biology. It is intended to help you prepare for **Unit F215: Control, Genomes and Environment**. It is divided into two sections:

- **Content Guidance** — here you will find key facts, key concepts and links with other parts of the AS/A2 biology course. **Focus on mathematical skills** will help with the chi-squared test and the Hardy–Weinberg principle. The **synoptic links** are intended to show you how topics in this unit build on information you learnt at AS. It also shows you how information in this unit links to topics in Unit F214: Communication, Homeostasis and Energy.
- **Questions and Answers** — here there are questions that cover all nine sections in this unit, with answers written by two students and examiner's comments.

This is not just a revision aid. This is a guide to the whole unit and you can use it throughout the A2 course. You should read other sources of information such as textbooks, articles from *Biological Sciences Review*, published by Philip Allan Updates, and websites. It is a good idea to use animations to follow the complex processes described in this guide. I have recommended websites that give up-to-date information about topics in this unit that are developing rapidly.

The **Content Guidance** section will help you to:

- understand the key concepts of each section
- organise your notes and to check that you have highlighted the important points (key facts) — little 'chunks' of knowledge that you can remember
- check that you understand the links to practical work, since you will need your knowledge of this unit when doing the practical tasks in Unit F216: Practical Skills in Biology 2
- understand how these little 'chunks' fit into the wider picture; this will help to support Units F214 and F216

You may be entered for F214 in January of your A2 year. If this is the case, you will not have time to study much, if any, of F215 before then. However, many students take the examinations for both units in June. This gives you more time to see the connections between topics in the two units.

The **Questions and Answers** section will help you to:

- check the way examiners ask questions at A2
- understand what the examiners mean by terms such as 'describe' and 'explain'
- interpret the question material, especially any data that the examiners give you
- write concisely and answer the questions that the examiners set

In this guide there are references to the three tasks you will carry out as part of Unit F216:

- the **qualitative task**, e.g. carrying out an experiment that does not give you anything to measure or determine — it may involve recording colours or drawing from a specimen or from a microscope slide
- the **quantitative task**, e.g. carrying out a practical task in which you record measurements

- the **evaluative task**, e.g. commenting critically on the practical procedure and the results you obtained in the quantitative task

There is a Student Unit Guide specifically for the practical assessment in F216. You will find some references to it in this guide.

A2 biology

The diagram below shows you the three units that make up the A2 course. You should have a copy of the specification for the whole of the course. Keep it in your file with your notes and refer to it constantly. You should know exactly which topics you have covered so far and how much more you have to do.

Unit F214 Communication, Homeostasis and Energy	+	**Unit F215** Control, Genomes and Environment	+	**Unit F216** Practical Skills in Biology 2

The specification outlines what you are expected to learn and do. The content of the specification is written as **learning outcomes**; these state what you should be able to do after studying and revising each topic. Some learning outcomes are very precise and cover just a small amount of factual information. Some are much broader. Do not think that any two learning outcomes will take exactly the same length of time to cover in class or during revision. It is a good idea to write a glossary for the words in the learning outcomes; the examiners will expect you to know what they mean. This guide should help you to do this.

Synoptic introduction to Unit F215

Approximately 20% of the marks in the unit test are synoptic. This means that you are expected to use your knowledge and understanding of the two AS units and Unit F214 when answering questions. To help you with this, the links to these other units are made clear throughout the guide. Several aspects of biology are covered in some detail in F215: structure and function of genes, genetics, variation and selection, cloning, biotechnology and genetic engineering, ecosystems and populations, plant and animal responses. All of these link to the AS course and to Unit F214 and you should revise appropriate topics before you start each section. The most important are:

- mitosis (Unit F211)
- structure of nucleic acids (Unit F212)
- enzymes (Unit F212)
- biodiversity, classification, evolution and maintaining biodiversity (Unit F212)
- nerves and hormones (Unit F214)
- respiration (Unit F214)

Examiners have to set some challenging questions in the unit test. Identifying the main themes in this unit and making links between different topics are two ways in which you can prepare for these questions. You should also try the numerical and other problems in the knowledge checks throughout this guide.

Content Guidance

Cellular control and variation

Key concepts you must understand

In Unit F211 you learnt about protein synthesis and in Unit F212 you learnt about the structure of nucleic acids and proteins. During **semi-conservative replication**, DNA polynucleotides act as **templates** for the assembly of new polynucleotides using the four deoxyribonucleoside triphosphates (dNTPs) with the four bases adenine, cytosine, guanine and thymine. Bases on the template polynucleotide determine which nucleotide is next in the sequence — remember that in DNA cytosine always pairs with guanine and adenine always pairs with thymine. Replication occurs during interphase of the cell cycle, as does transcription in which DNA acts as a template for the assembly of RNA nucleotides to make mRNA for protein synthesis.

Examiner tip
Before you read this section on cellular control and variation, you should revise what you learned in Units F211 and F212 on the cell cycle, mitosis, replication and protein synthesis. Above all you should revise the structure of nucleic acids, DNA and RNA.

Key facts you must know

One gene: one polypeptide

A gene determines the sequence of amino acids in a polypeptide, for example enzymes, α- and β-globins of haemoglobin, cell surface antigens, receptors and hormones. Many proteins are composed of several polypeptides and have quaternary structure, for example:

- a molecule of catalase is made of four identical polypeptides
- a molecule of adult haemoglobin is made of two α-globin and two β-globin chains

Examiner tip
A nucleoside is a compound composed of a base (A, T, C or G) plus a pentose sugar. In a deoxyribonucleoside the sugar is deoxyribose. Use the term polynucleotide rather than 'chain' when writing about DNA.

Examiner tip
Quaternary structure applies if a protein has more than one polypeptide in its structure. Molecules of haemoglobin and catalase have four polypeptides each; many proteins have two or three.

The four bases (A, T, C and G) code for 20 different amino acids. Some examples are given in Table 1.

Table 1 Genetic dictionary: DNA triplet codes for three amino acids on the coding and template polynucleotides

Amino acid	Triplets on coding polynucleotide of DNA	Triplets on template polynucleotide of DNA
Glycine	GGT; GGG; GGC; GGA	CCA; CCC; CCG; CCT
Valine	GTT; GTG; GTC; GTA	CAA; CAC; CAG; CAT
Cysteine	TGT; TGC	ACA; ACG

The genetic code is described as **degenerate** because there are more codes than are needed for each amino acid. This is an advantage because the codes for each amino acid usually differ in the third base of the triplet, which reduces the effects of mutation. There are three 'stop' codes that do not code for amino acids. They indicate the end of a sequence (e.g. TGA on the coding strand of DNA). The triplet code for the amino acids is called the **genetic code** and is shown in a genetic dictionary as the RNA code or as one of the DNA codes. Each group of three bases in mRNA that codes for an amino acid is known as a **codon**.

Knowledge check 1
State the DNA triplets for histidine (his) and phenylalanine (phe).

Knowledge check 2

Explain why the code for amino acids must be a triplet code and not a two-base code.

Examiner tip

You have not read much of this book and already there are many new terms to learn. Write your own glossary as you read this book using other sources to help you. You will see how useful this is when you read the Questions and Answers section of this guide.

The complete genetic code can be found on many websites. You are not expected to remember any of the triplets or codons, but you may have to use the genetic code in answering a question, in which case it will be provided for you.

Protein synthesis

The three stages in protein synthesis are transcription, amino acid activation and translation, which are best followed in an animation:

- www.johnkyrk.com (DNA transcription and DNA translation)
- http://highered.mcgraw-hill.com/sites/0072437316/student_view0/chapter15/animations.html (protein synthesis)

Transcription

β-cells in the islets of Langerhans in the pancreas make insulin. It is only in these cells that the gene for insulin is switched on. Each β-cell has two copies of this gene, but many copies are needed to send to the thousands of ribosomes in the cell to make the quantities of insulin required. Short-lived copies of the gene are made by transcription. The copies are molecules of mRNA. The process of transcription is shown in Figure 1. Note that the base sequence of the mRNA is the same as that of the coding polynucleotide and complementary to that of the template polynucleotide (U replaces T in RNA).

Examiner tip

You should recall the role of β-cells in the pancreas from F214. See page 33 of the Unit Guide to F214. Questions on protein synthesis set in Unit F215 could be based on proteins you have studied in other Units.

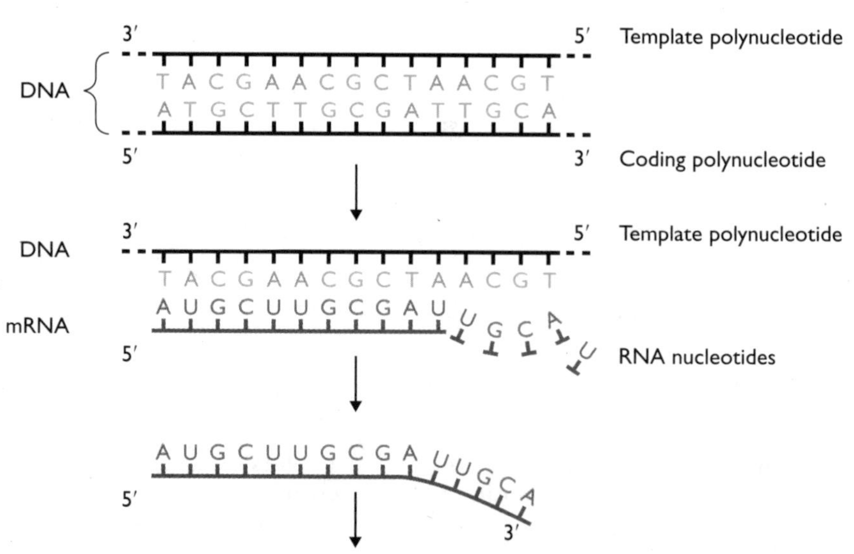

Figure 1 Transcription of the template polynucleotide in DNA

(1) Hydrogen bonds between the bases in the two polynucleotide chains break in the area of DNA that corresponds to the insulin gene.

(2) One polynucleotide acts as a template for the synthesis of mRNA.

(3) Free RNA nucleotides in the nucleus pair up with the exposed bases on the template polynucleotide.

(4) The nucleotides are joined together to form a polynucleotide — mRNA. This process is catalysed by the enzyme RNA polymerase.

(5) mRNA leaves the nucleus through a nuclear pore.

Knowledge check 3

List the first five amino acids in the polypeptide coded for by the gene in Figure 1.

Knowledge check 4

What are the likely effects of reading errors in transcription?

Knowledge check 5

State what 3' and 5' signify.

Amino acid activation

Amino acids are 'identified' or 'tagged' by combining them with molecules of transfer RNA (tRNA). You could think of the nucleotide 'labels' as being similar to bar codes. The nucleotide 'bar code' is the **anticodon**. Enzymes in the cytoplasm attach amino acids to specific tRNA molecules. This is not a random process — each amino acid is identified by a specific tRNA molecule (Figure 2). The hydrolysis of ATP provides the energy for the process.

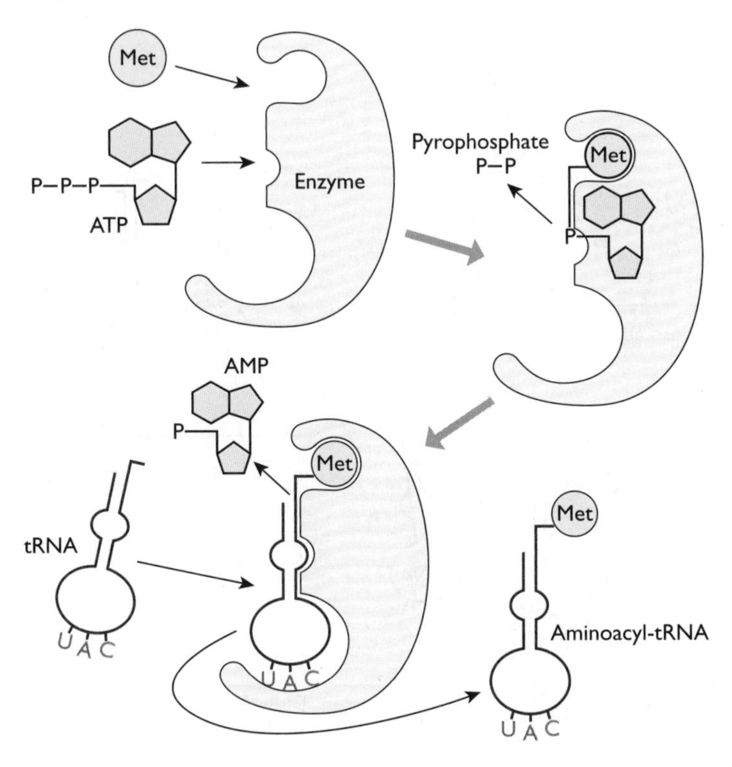

Figure 2 The enzyme shown here only accepts methionine and its specific tRNA molecule

Translation

(1) The mRNA molecule joins with a ribosome in the cytoplasm.
(2) Each ribosome has two sites to hold two tRNA molecules at the same time. Each tRNA molecule is attached to an amino acid.
(3) Each tRNA molecule has a sequence of three bases (anticodon) that pairs with three bases (codon) on mRNA.
(4) tRNA and mRNA pair together following the rules of complementary base pairing (A with U; C with G).
(5) A condensation reaction occurs between the amino acids to form a peptide bond.
(6) The ribosome moves along the mRNA molecule 'reading' the sequence of bases.
(7) As this happens, a polypeptide grows by the addition of new amino acid molecules.
(8) When an amino acid has joined to the growing chain, its tRNA molecule leaves the ribosome to attach to another amino acid.

(9) When the ribosome reaches a stop codon, the polypeptide breaks away and begins to fold spontaneously into its secondary and tertiary structure.

(10) The cell processes the polypeptide, perhaps by combining it with other polypeptides to form a protein with quaternary structure, as happens in the formation of haemoglobin. In β-cells in the islets of Langerhans in the pancreas a long polypeptide is cut up into two smaller polypeptides that are joined together by disulfide bonds to give insulin its quaternary structure.

Examiner tip
Lysozyme is an enzyme made of a single polypeptide. Search online for images of its tertiary structure and compare with the structure of haemoglobin or catalase.

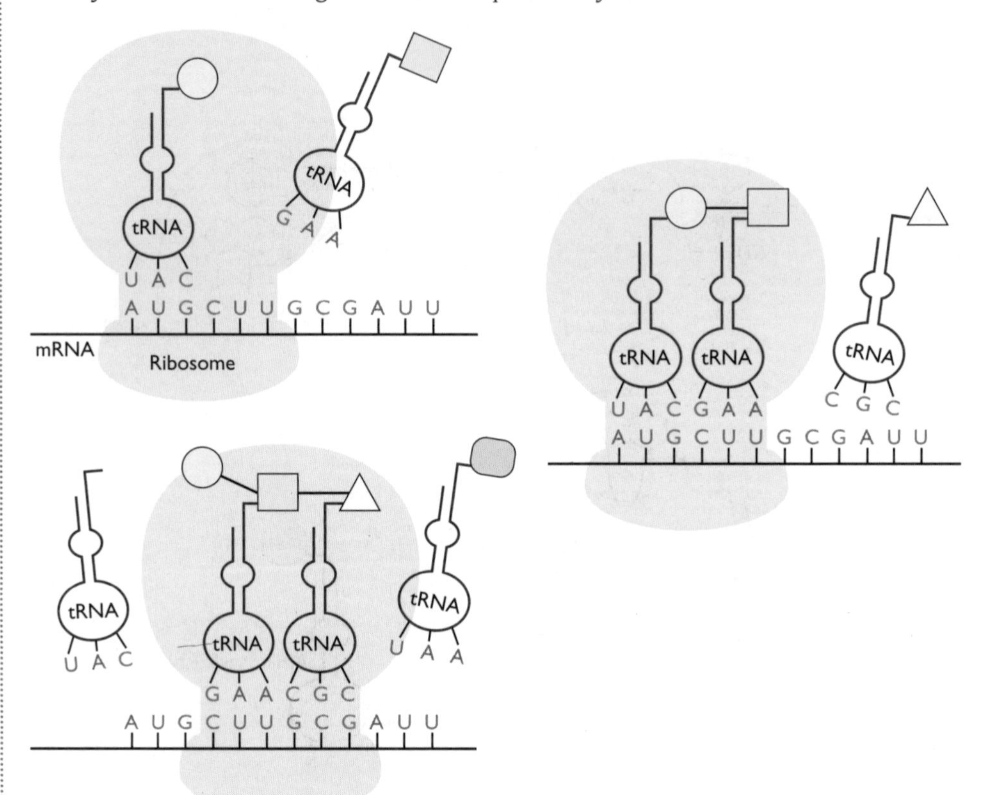

Figure 3 Translation

Mutation

Proofreading by DNA polymerase during replication means that mistakes are rare events and base sequences of genes remain the same generation after generation. However, mistakes do happen and these lead to mutations. **Gene mutations** involve relatively small changes to the sequence of nucleotides in DNA. **Mutagens**, such as radiation (e.g. X-rays) and certain chemicals (e.g. benzpyrene in tobacco smoke) increase the chances of mutation.

Types of gene mutation

Substitution A base pair changes, e.g. from A–T to C–G. If this happens in the first base pair of a triplet then it is likely to change one amino acid in the primary structure of a polypeptide. Changes to the second or third base are less likely to change the amino acid and are known as 'silent' mutations.

Frameshift A deletion or insertion of a base pair changes the 'reading frame' during translation so the amino acid sequence downstream of the mutation changes (see Figure 4). Depending on where in the gene this happens it can have catastrophic effects on the polypeptide so that it does not work, or works only poorly. The polypeptide may be longer or shorter than the original if the effect of the mutation is to remove or add a stop codon. Many recessive mutations which do not code for functioning proteins are frameshifts.

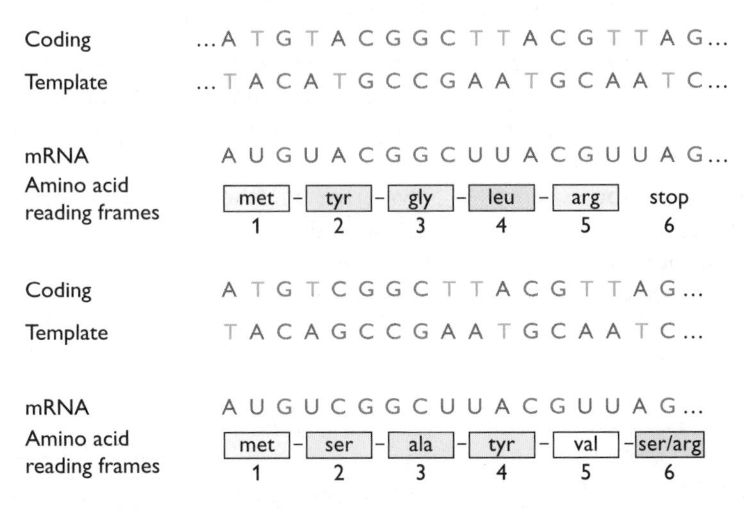

Figure 4 The effects of a frameshift mutation on a sequence of five amino acids. The fifth base pair has been deleted

Knowledge check 6

Summarise the effects of the deletion of the fifth base pair as shown in Figure 4.

Knowledge check 7

What are the effects of an insertion of the base pair A–T after the first triplet in the DNA in Figure 4? (A is on the coding polynucleotide.)

Stutter mutations also occur where triplets are repeated. This is called a 'stutter'. The rare neurological disease Huntington's disorder is caused by a genetic stutter.

Neutral mutations have no effect on the fitness or survival of the organism. They are of different types:

- The mutant triplet codes for the same amino acid, e.g. CCG and CCT ('silent' mutation).
- The triplet codes for a different amino acid, but this makes no difference to the function of the polypeptide.
- The mutant polypeptide functions in a different way but this does not give an advantage or a disadvantage to the organism.
- The mutation occurs in non-coding regions of DNA and is not expressed in the phenotype.

Knowledge check 8

There are genes that code for tRNA molecules. What would be the effect of a mutation in the DNA that codes for the anticodon region of a tRNA molecule?

Some mutations are always harmful because they destroy the function of the protein. Others may have effects on the phenotype that confer no advantage and may even be disadvantageous. However, environments may change and selection acts on phenotypes not genes, so a feature that is neutral or disadvantageous may become advantageous in the future. This is why genetic variation is so important (see p. 15).

Genes at work

Some genes are transcribed and translated all the time — for example, genes for **constitutive** enzymes, such as those that catalyse stages in glycolysis. Other enzymes are made only when they are required. It would be a waste of energy to make these in the absence of a substrate for them to catalyse. One of these **inducible** enzymes is β-galactosidase, which hydrolyses lactose in the bacterium *Escherichia coli*. The production of the enzyme is controlled along with two other proteins that are required for the absorption and metabolism of lactose. The three genes are transcribed together giving one mRNA transcript.

Examiner tip

Working through this site will help with the *How Science Works* theme:

www.dnaftb.org/33/

It is a good idea to follow an animation as you read about the *lac* operon:

http://highered.mcgraw-hill.com/olc/dl/120080/bio27.swf

Figure 5 shows how these genes are turned off when there is no lactose in the surroundings. However, when lactose is present it enters *E. coli* and acts to turn on the genes so they are transcribed and the three proteins are produced. The **operon** consists of the following:

- a **promoter region**, where RNA polymerase binds to start transcription
- an **operator region**, where an inhibitor binds
- **structural genes** for β-galactosidase, a membrane transport protein for lactose and another enzyme

Knowledge check 9

Explain how the *lac* operon is controlled when there is (a) glucose present, but no lactose, and (b) glucose and lactose both present.

Knowledge check 10

If there is no glucose and no lactose in the medium, β-galactosidase is not produced. Explain how the *lac* operon is controlled in these conditions.

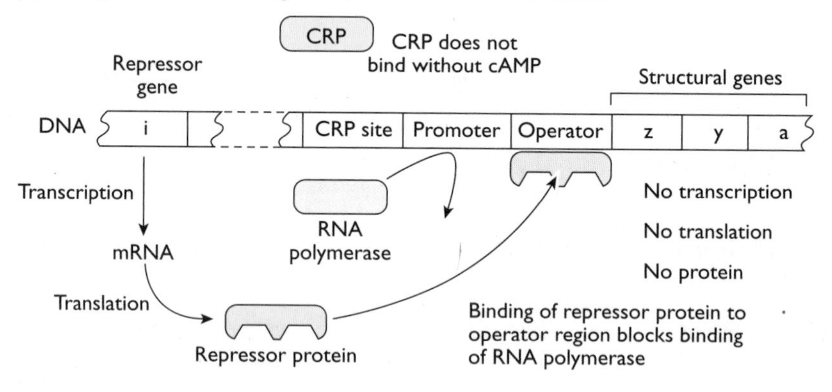

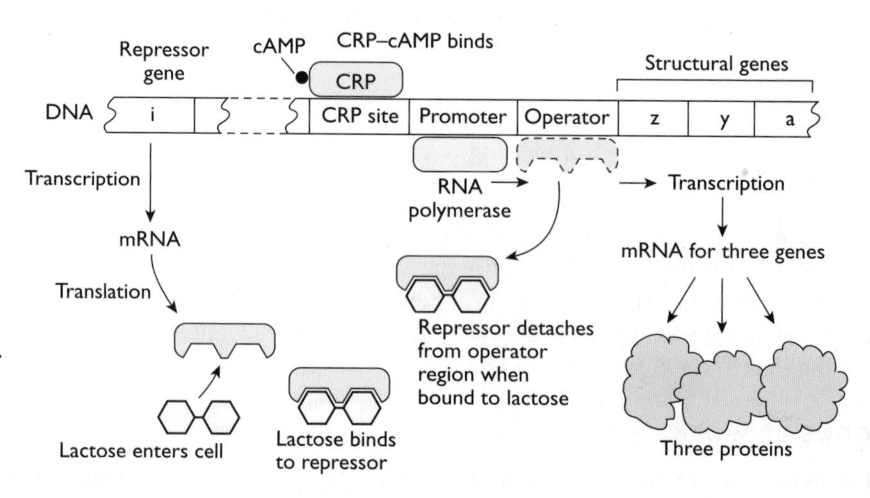

Figure 5 How the *lac* operon functions (a) when lactose is absent and glucose is present in high concentration and (b) when lactose is present and there is little or no glucose

In addition to the *lac* operon, there is a **regulator gene** elsewhere on the bacterial chromosome that codes for a repressor substance that binds to the promoter region, inhibiting transcription of the three *lac* genes **z**, **y** and **a**.

Control by the repressor is **negative control** — lactose 'turns on' the *lac* operon by inhibiting the repressor, which is itself an inhibitor of transcription. However, this happens only when there is no glucose present. RNA polymerase does not bind readily to the promoter region. A protein known as CRP (cyclic AMP receptor protein) helps the binding of RNA polymerase, which transcribes the genes only when CRP is bound to DNA. If both glucose and lactose are present, then the operon is not switched on because it is more efficient for *E. coli* to respire glucose. If glucose is present, there is a low concentration of the regulatory substance, cyclic AMP (cAMP). In the absence of glucose, the concentration of cAMP increases and it combines with CRP. This binding changes the three-dimensional shape of CRP and exposes its DNA-binding site. Once bound to DNA, the activated CRP helps RNA polymerase to bind to the promoter region to start transcription. This action by CRP is **positive control**.

Operons are found in both prokaryotes and eukaryotes, although many eukaryotic genes are controlled differently and transcribed individually, rather than in groups.

Genes in development

Cells differentiate and become specialised for specific roles. To do this, they switch some genes on and some off in a carefully organised sequence. **Homeobox** genes code for transcription factors that bind to DNA. The homeobox is a sequence of bases in DNA that codes for a region of 60 amino acids in proteins that bind to DNA to regulate transcription by turning genes on and off in correct patterns. These genes control the early development of animals, plants and fungi. They give the basic pattern to the body — for example, they control the segmentation pattern of insects and mammals and the development of wings and limbs (see Figure 6). The homeobox sequences of organisms are almost identical because they all have the same function — coding for transcription factors.

Apoptosis is programmed cell death. This happens throughout life, but is an important part of development. As the immune system develops, many potential T lymphocytes die because they carry cell-surface antigens that are complementary to proteins on the surface of our own cells and would destroy them. Cells between the fingers and toes are also destroyed. If not, we would have webbed hands and feet. Cell death occurs in the endometrium during menstruation and also during the formation of synaptic connections in the nervous system. Cells respond to external and internal signals that trigger an ordered sequence of changes in the cytoplasm ensuring that cells are removed efficiently without the release of hydrolytic enzymes that would damage surrounding tissue and cause inflammation. Signals stimulate production of proteins that promote apoptosis and these activate caspases that are part of an enzyme cascade that breaks down structural proteins, such as those of the cytoskeleton. Anti-apoptosis genes code for inhibitor proteins that prevent apoptosis in healthy cells. Apoptosis occurs in plants to remove cells infected with viruses.

Some Hox genes work by activating genes that promote apoptosis. The Hox gene known as deformed (**Dfd**), in *Drosophila* activates the gene known as reaper (**Rpr**) to induce cell death in order to separate the maxillary and mandibular head lobes.

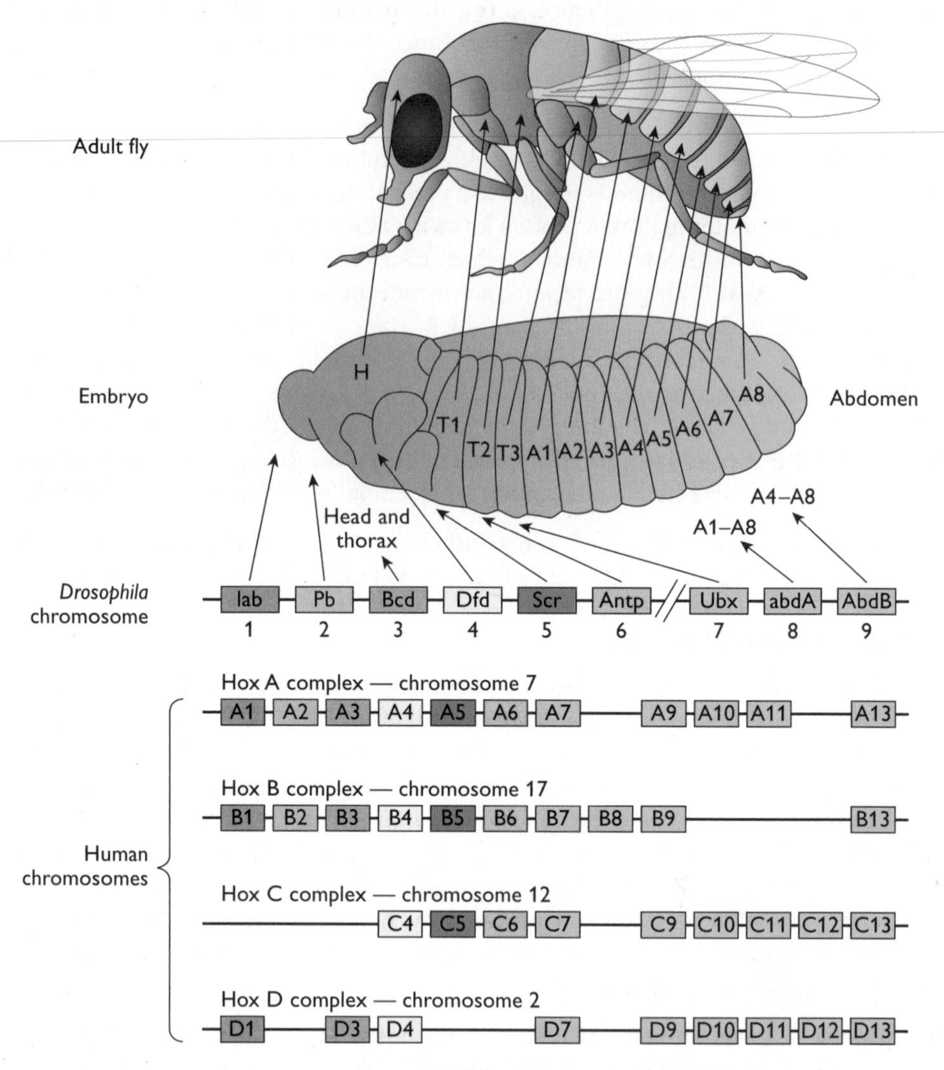

Figure 6 Nine homeobox genes in the fruit fly, *Drosophila melanogaster*, control the development of the embryo in sequence: head, thorax and abdomen. The same genes are duplicated and found in four Hox clusters in humans

Examiner tip
You do not need to know about the specific genes and their effects as shown in Figure 6, but expect them to be used as contexts in exam questions. Read about some of them and learn some examples. Search for infomation about antennapedia (*Antp*) in *Drosophila*.

Synoptic links

This is a good opportunity to revise the structure of prokaryotic and eukaryotic cells from Unit F211. You could be asked where the stages in protein synthesis occur in eukaryotic cells. Enzymes that already exist in the cytoplasm may be activated and inactivated by compounds such as cyclic AMP (cAMP). cAMP acts as a second messenger to activate a protein kinase enzyme at the start of a cascade in response to adrenaline and glucagon. cAMP fits into a site on the enzyme and changes its tertiary structure. The enzyme becomes active, catalysing the phosphorylation of another enzyme in the cascade. The role of cAMP as a second messenger is covered in Unit F214 (see p. 30 in the guide for Unit F214).

Summary

- Genes code for the assembly of amino acids to make polypeptides, such as enzymes. Some enzymes, such as lysozyme, are composed of a single polypeptide; catalase is composed of four.
- The genetic code is the sequence of the four bases in DNA (A, T, C and G) that code for amino acids. There are 61 different triplets in DNA and codons in RNA that code for the 20 amino acids. There are three stop codons that do not code for any amino acids.
- Transcription is the copying of a nucleotide sequence in DNA into a complementary sequence in mRNA. Translation is the assembly of amino acids on ribosomes using the sequence of codons to specify the sequence of amino acids in a polypeptide.
- Transfer RNA molecules are activated by combining with specific amino acids. The anticodons on tRNA molecules pair with codons on mRNA so that amino acids are assembled in the correct sequence. Peptide bonds form between amino acids.
- Gene mutations change the sequence of nucleotides in DNA molecules. Substitution mutations result in a change to a single amino acid in a polypeptide; frameshift mutations result in a change to all the amino acids downstream from the site of the mutation (deletion or addition).
- Most mutations have a harmful effect on the phenotype; some substitution mutations can have beneficial effects on the phenotype. Neutral mutations have no effect on the phenotype.
- Cyclic AMP is a second messenger that combines with proteins to activate or inactivate them by altering their three-dimensional structure.
- The *lac* operon is responsible for the control of the production of three proteins in the prokaryote, *Escherichia coli*.
- The genes that control the development of body plans are similar in plants, animals and fungi. The homeobox sequences are sequences of nucleotides in these genes that code for transcription factors that bind to DNA. Their sequences are very similar because they code for parts of proteins that have the same shape for binding to DNA.
- Apoptosis is programmed cell death that changes body plans, such as removing tissue between digits (fingers and toes).

Meiosis and variation

Key concepts you must understand

Variation is the sum of the differences between species and within species. In this section we are interested in **intraspecific variation** — variation within a species — and how it is generated.

Phenotypic variation is variation that can either be seen (e.g. colour of fur or feathers) or detected by methods such as using antibodies to identify blood groups and electrophoresis to show variation in proteins. Genotypic variation is variation in the genes and refers to the alleles of different genes in an individual.

New genetic material is generated by mutation, but without meiosis it is unlikely that much of this variation would be expressed. Variation is further increased by random mating between individuals and by random fertilisation.

Key facts you must know

There are two forms of variation in phenotypic features. In **continuous variation** there is a range of variation between two extremes; in **discontinuous variation** there are distinct categories with no range of intermediates between them. The two forms of variation are compared in Table 2.

The ABO blood group system and sickle-cell anaemia are examples of discontinuous variation (see Table 3).

Table 3 shows that in the ABO system there are three alleles. This is an example of multiple alleles (more than two).

Table 2 Comparison between continuous and discontinuous variation

Feature	Continuous variation	Discontinuous variation
Appearance of phenotype	Quantitative, e.g. mass and length, with many intermediates	Qualitative, e.g. presence or absence of a feature with no intermediates
Number of genes	Many (polygenic)	Few, often just one gene (monogenic)
Effects of genes	All have the same effect	Different effects
Effects of alleles at each locus	Small effects	Large effects
Effect of the environment	Large	Small or non-existent
Representation	Frequency histogram	Bar chart
Examples: Microorganisms	Quantitative feature: hyphal length in filamentous fungi, e.g. *Penicillium chrysogenum*	Qualitative feature: antibiotic resistant and susceptible strains of bacteria
Plants	Mass, length and width of leaves; height of shoots, length of roots	Tall and dwarf pea plants; flower colour; position of flowers on plant
Animals	Mass, height, length of animals	ABO, Rhesus and MN blood groups in primates; attached and free ear lobes in humans

Table 3 Comparison between the ABO blood group system and sickle-cell anaemia

Feature	ABO blood group system	Sickle-cell anaemia
Location of gene	Chromosome 9	Chromosome 11
Gene	Codes for an enzyme that adds a sugar molecule to a cell-surface antigen	Codes for the β-globin polypeptide in haemoglobin
Number of alleles	Three ($\mathbf{I^A}$, $\mathbf{I^B}$, $\mathbf{I^O}$)	Two ($\mathbf{Hb^A}$, $\mathbf{Hb^S}$)
Number of genotypes	Six	Three
Number of different phenotypes	Four (A, B, AB and O)	Two or three (normal; sickle-cell trait; sickle-cell disease)

Examiner tip

Remember when studying Table 3 that any one individual can only have two alleles because he or she is diploid.

Knowledge check 11

Justify classifying the alleles $\mathbf{Hb^A}$ and $\mathbf{Hb^S}$ as codominant.

Knowledge check 12

Write out all six genotypes for the ABO system.

Knowledge check 13

Explain what AB+ means.

Genotype and environment contribute to phenotype

Milk yield is an example of continuous variation. It is influenced by the environment and by genes. In the UK, milk yield has increased significantly over the past 60 years from an average of 3000 dm^3 per cow per year in the 1940s to an estimated 7406 dm^3 per cow per year in 2010–11. The volume of milk produced by a cow is determined by the genes inherited from her parents and by environmental factors, such as quality and quantity of food. Milk yield is controlled by many genes each with many alleles. Each allele has an additive effect. Milk yield is an example of a sex-limited condition as it is expressed in one sex only. The male parent has genes for milk yield even though they are never expressed in him. The improvement in milk yield is the result of selective breeding, improved feeding and changes in the management of dairy herds (see p. 32 for more on this topic).

Without variation there can be no selection. Variation provides the 'raw material' for selection.

Role of meiosis in variation

The specification talks about 'the behaviour of chromosomes'. It is important that you understand the structure of chromosomes and what happens to them during the mitotic cell cycle before reading about meiosis and interpreting photographs and diagrams.

Examiner tip

Before you start this section, revise thoroughly the events that occur in the mitotic cell cycle.

The only way in which completely new genetic material is generated is by gene mutation. In most organisms, meiosis occurs before the fusion of gametes — most often during gamete formation. In some organisms, for example fungi, meiosis occurs after the fusion of gametes. Whenever meiosis occurs in the life cycle it leads to new combinations of the alleles of all the genes in an organism. Variation is the result of these events during meiosis:

- gene mutation in gamete-forming (germ-line) cells
- chromosome mutation (change in structure or number of chromosomes)
- segregation of allelic pairs in meiosis I
- independent assortment of maternal and paternal chromosomes during meiosis I
- independent assortment of **sister chromatids** during meiosis II
- crossing-over between **non-sister chromatids** during meiosis I
- halving the chromosome number so that further variation occurs at fertilisation when new combinations of alleles may occur

Knowledge check 14

Define the term chromatid and distinguish between sister chromatids and non-sister chromatids.

Without meiosis the chromosome number would double with each generation. Meiosis occurs during the formation of gametes (sperm and eggs) in animals (see Figure 7) and during formation of spores in plants.

You should look at microscope slides of pollen formation in the lily, *Lilium*. Colour photographs of these stages are at: www.iasprr.org/old/iasprr-pix/lily/

You should then compare slides and photographs showing the stages of meiosis with diagrams and make sure that you can recognise the stages.

There are animations of meiosis at:

www.johnkyrk.com/meiosis.html

http://highered.mcgraw-hill.com/sites/0072437316/student_view0/chapter12/animations.html

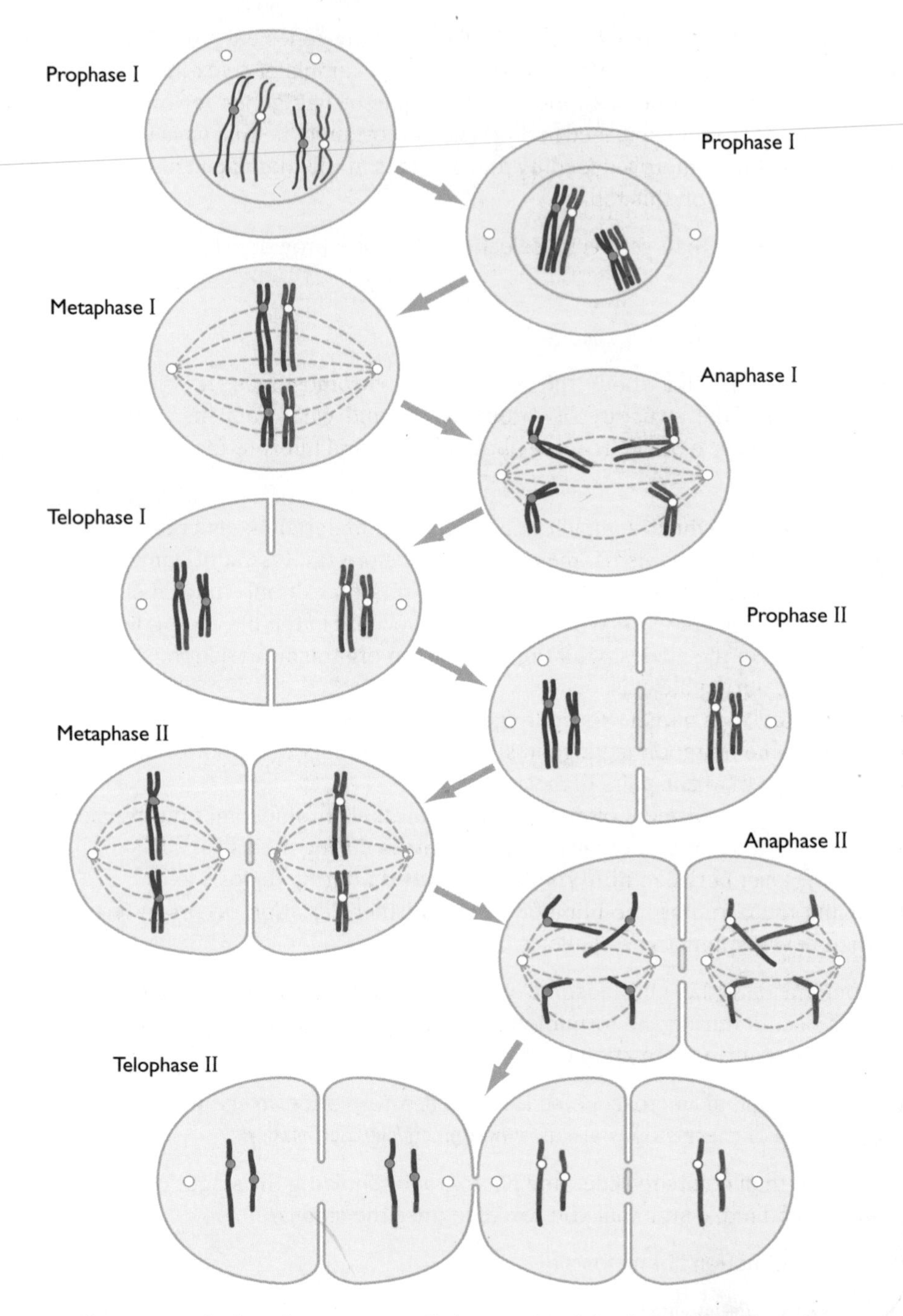

Figure 7 Meiosis to form sperm cells in an animal that has a diploid number of 4. Maternal chromosomes are blue, paternal chromosomes are red

Prophase I

- Chromosomes condense so that they become shorter and thicker, and visible in the light microscope
- Homologous chromosomes pair to form bivalents (maternal chromosomes are blue, paternal are red)

- Chiasmata (singular: chiasma) form to hold chromosomes together; non-sister chromatids join, break and exchange parts in crossing over
- Nuclear membrane breaks up into small sacs of membrane which become part of the endoplasmic reticulum; centrioles replicate and move to opposite poles and form the spindle

Metaphase I
- Bivalents move to the equatorial (or metaphase) plate across the centre of the cell
- Paternal and maternal chromosomes in each bivalent position themselves independently of the others
- Microtubules attach to the centromere of each chromosome

Anaphase I

- Chromosomes (each with two chromatids) are pulled towards the poles by the shortening of the microtubules

Telophase I
- Chromosomes reach opposite poles
- Nuclear membranes reform to make two daughter nuclei that have half the number of chromosomes of the parent cell — these nuclei are **haploid**
- Cytokinesis occurs — the cell surface membrane 'pinches in' leaving small cytoplasmic bridges between the cells

An interphase may occur between the divisions of meiosis I and II, in which case the chromosomes uncoil.

Prophase II
- Centrioles replicate and move to poles that are at right angles to those in meiosis I
- Nuclear membranes break up

Metaphase II
- Individual chromosomes align on the equator with their chromatids randomly arranged (important if crossing over has occurred in meiosis I)
- Microtubules attach to the centromeres

Anaphase II

- Sister chromatids break apart at the centromere and move to opposite poles

Telophase II
- Nuclear membranes reform
- Cells divide to give four haploid cells that are genetically different from one another and from the parent cell

The haploid cells produced differentiate into sperm cells.

During prophase and metaphase of the first stage of meiosis chromosomes are 'double-stranded' as they have two chromatids joined together at the centromere. These are produced during interphase by replication. During anaphase I the chromatids separate to form two 'single-stranded' chromosomes.

Examiner tip

Example 29 in the Unit guide to F213/F216 is about the products of meiosis and includes data that could be set in the examination on Unit F215.

Examiner tip

Common questions on meiosis involve sequencing the stages. To prepare for these make some drawings of the stages of meiosis and find some photomicrographs. Cut them up and put them in the correct sequence using Figure 7 to help you. Then repeat without using the figure.

Examiner tip

Notice how the terms gene and allele are used in this section. Gene refers to a length of DNA that codes for a particular polypeptide. Alleles are different versions of the gene with different base sequences: differing in one base pair (e.g. sickle cell anaemia) or more.

Examiner tip

See p. 72 for some information about online resources to help you with understanding the genetics of *Drosophila*.

Knowledge check 15

Meiosis and fertilisation are two causes of variation within a species. Name another cause of variation.

Meiosis, genes, alleles and variation

A genetic diagram, such as that in Figure 8, shows how genes are inherited. The diagram shows a monohybrid cross involving the inheritance of the gene for wing length in *D. melanogaster*. There are two alleles: long wing (also known as wild type) (**W**) and vestigial (**w**). Vestigial means very small. During meiosis in the F_1 flies the alleles separate because they are on homologous chromosomes that separate during meiosis I. This separation is called **segregation of allelic pairs**. Notice also that fertilisation gives rise to variation since gametes with the allele **W** can fuse with gametes with **W** or **w** to give flies with the homozygous dominant genotype (**WW**), flies that are heterozygous (**Ww**) and flies that are homozygous recessive (**ww**).

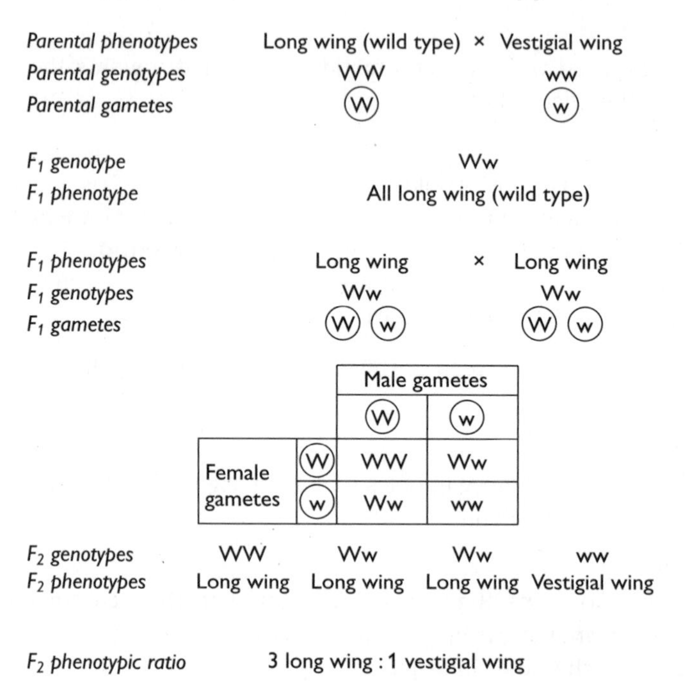

Figure 8 A monohybrid cross: the inheritance of the gene for wing length in *D. melanogaster*

The chequerboard or Punnett square (Figure 8) is the best way to show the genotypes of the next generation, even when there are few genotypes involved. Note that the Punnett square shows the possible outcomes — the genotypes do not represent actual organisms. It shows the probabilities of different genotypes and phenotypes in the next generation.

Figure 9 shows what happens in the inheritance of two **unlinked genes** (not on the same chromosome). Figure 10 shows the arrangement of maternal and paternal chromosomes at metaphase in meiosis I and how this is responsible for the production of four types of gamete. There is a 50% chance that a cell undergoing meiosis will be like **A** in Figure 10 or like **B**. This gives a 25% chance of forming each of the four gametes with different genotypes. As a result of the random assortment of homologous chromosomes at the equator during metaphase I, the allelic pairs, **W/w** (wing length) and **E/e** (body colour), segregate independently of one another and the gametes receive a mixture of maternal and paternal chromosomes (see Figure 10).

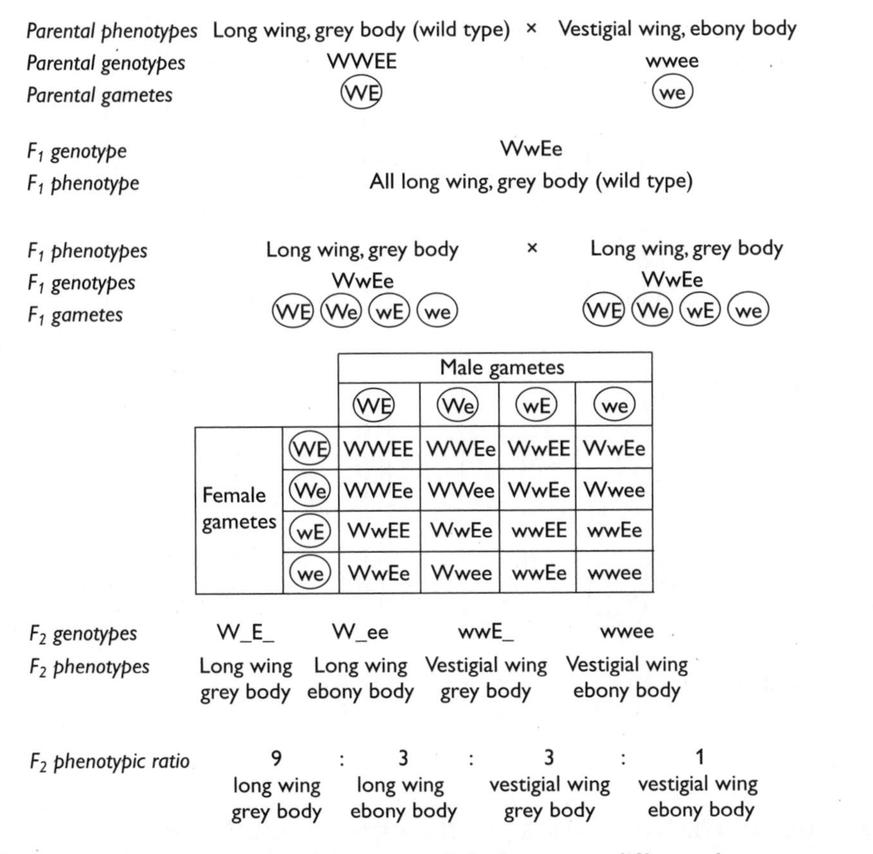

		Male gametes			
		WE	We	wE	we
Female gametes	WE	WWEE	WWEe	WwEE	WwEe
	We	WWEe	WWee	WwEe	Wwee
	wE	WwEE	WwEe	wwEE	wwEe
	we	WwEe	Wwee	wwEe	wwee

Figure 9 A dihybrid cross involving two unlinked genes on different chromosomes

Examiner tip
In dihybrid crosses with unlinked genes, write the genotypes of diploid organisms as shown here, e.g. WwEE, not WEwE.

Examiner tip
The dash (_) in Figure 9 represents either allele for each gene locus, **W/w** and **E/e**. It is useful to use dashes when completing genetic diagrams; if you do this in an exam question you should explain what it means.

Codominance

Sometimes there is no dominance between two (or more) alleles at the same locus. Both alleles are expressed in the phenotype of a heterozygote. The M and N blood groups show this. Two alleles code for glycoprotein antigens (M and N) on the surface of red blood cells. The genotypes and phenotypes are as follows:

- $GYPA^M\ GYPA^M$ — glycoprotein M; blood group M
- $GYPA^M\ GYPA^N$ — glycoproteins M and N; blood group MN
- $GYPA^N\ GYPA^N$ — glycoprotein N; blood group N

Other examples of codominance are coat colour in cattle (red, white and roan), flower colour in snapdragons (*Antirrhinum*) and the ABO blood group system in primates (see p. 16). These are all examples of discontinuous variation.

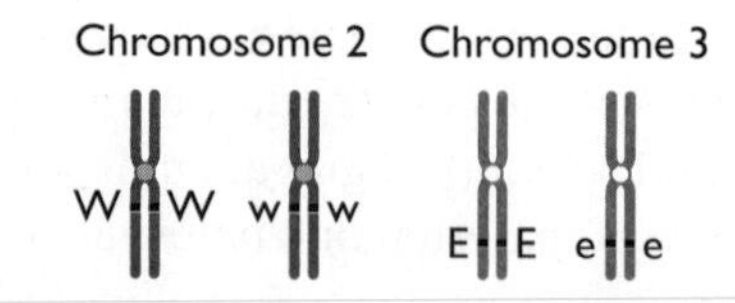

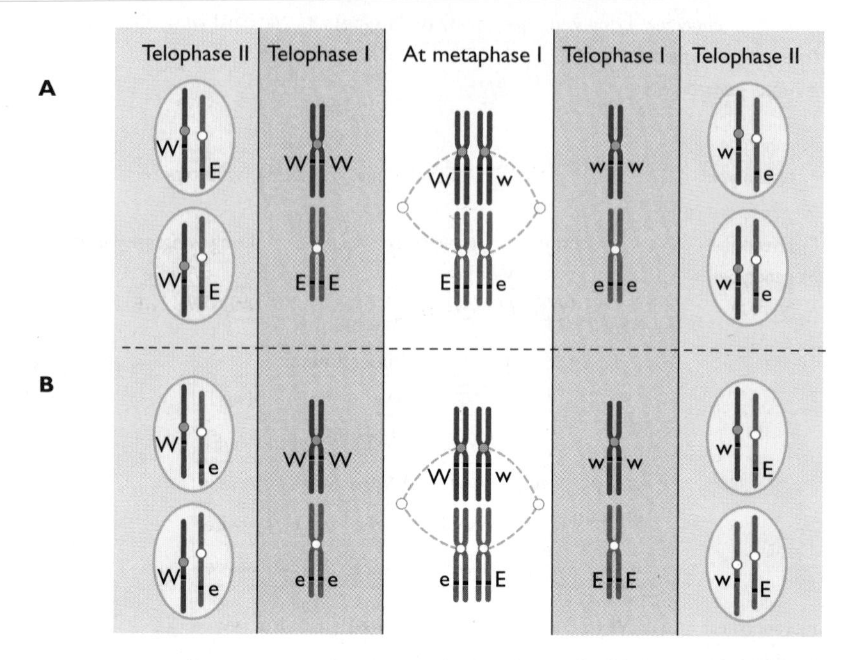

Figure 10 A and B show the two arrangements of maternal and paternal chromosomes in metaphase I responsible for independent assortment during meiosis in the formation of gametes in the F_1 generation

Focus on mathematical skills: chi-squared test

The chi-squared (χ^2) test is a statistical test with categoric data that is used to see if there is a difference between observed results and those predicted from theory. It gives an estimate of the probability that the difference is due to chance effects. The best way to understand the test is to follow an example.

Fruit flies of the F_1 generation from the cross in Figure 9 were crossed with fruit flies that were homozygous recessive for both loci. This is a **test cross** as the offspring show us what has happened in the formation of gametes in the individuals with the dominant phenotype and an 'unknown' genotype (here the F_1 generation). The results in the different categories were:

long wings grey body	vestigial wings grey body	long wings ebony body	vestigial wings ebony body
70	91	86	77

The ratio of phenotypes expected in the test cross offspring of a dihybrid cross such as this is 1:1:1:1, if we assume that the genes are not linked and independent assortment in the F_1 flies has occurred (see Figure 10).

Examiner tip

Make a genetic diagram of a dihybrid test cross to check that the ratio is 1:1:1:1.

The **chi-squared** (χ^2) test, and the probabilities shown in Table 4 on p. 24), are used to find out if these results differ significantly from the expected results or whether any differences are simply due to chance effects.

The **null hypothesis** states: there is no difference between the observed and expected results. The expected results are based on the theory about the inheritance pattern of the gene(s) concerned.

$$\chi^2 = \sum \frac{(O - E)^2}{E}$$

where Σ = sum of..., O = observed value and E = expected value

Categories	O	E	O − E	(O − E)^2	(O − E)^2/E
Long wings Grey body	70	81	−11	121	1.49
Vestigial wings Grey body	91	81	10	100	1.23
Long wings Ebony body	86	81	5	25	0.31
Vestigial wings Ebony body	77	81	−4	16	0.20
Totals	324	324		χ^2 =	3.23

Note that the differences between observed and expected results are squared to remove the negative signs, making all figures positive. Each of these figures is divided by the expected number to take into account the numbers of individuals.

The next step is to calculate the degrees of freedom (df). Imagine sorting through the offspring of the test cross. Once you have identified a fly and put it into one of the categories, how many other categories remain? That number is the degrees of freedom. It is calculated as the number of categories minus 1. In this example, the answer is 3. Read across the table of probabilities in Table 4 at df = 3 and find the **critical value**, which is 7.82. This is the value of χ^2 where there is a probability of 0.05 (5%) that the results could have been obtained by chance (5% is an arbitrarily agreed figure used by researchers). Where is 3.23?

Remember that > means greater than; < means less than. Shading the column headed $p = 0.05$ helps to find out if the χ^2 value is significant or not. See p. 81 for more about this.

p is less than 0.5 and greater than 0.1. The probability of getting this result is therefore between 10% and 50% which means that the result is due to chance effects such as random fertilisation. The difference is not statistically significant and so the null hypothesis can be accepted. If the value for χ^2 is greater than the critical value then the probability is less than 0.05 and there is a **significant difference** between observed and expected. If so, the prediction is rejected or refined or the experimental procedure is reviewed to see if there are any errors.

Table 4 Table of probabilities for χ^2

Degrees of freedom	Distribution of χ^2							
	← Increasing values of *p*			Probability, *p*				Decreasing values of *p* →
	0.99	**0.90**	**0.50**	**0.10**	**0.05**	**0.02**	**0.01**	**0.001**
1	0.00016	0.016	0.46	2.71	3.84	5.41	6.64	10.83
2	0.02	0.10	1.39	4.61	5.99	7.82	9.21	13.82
3	0.12	0.58	2.37	6.25	7.82	9.84	11.35	16.27
4	0.30	1.06	3.36	7.78	9.49	11.67	13.28	18.47

$p > 0.90$	$p > 0.05$	$p < 0.05$	$p < 0.01$	$p < 0.001$
Result is 'dodgy' = too good!	Result is not significantly different from expected outcome	Result is significantly different from expected outcome	Highly significant	Very highly significant

The table of probabilities that is used in examination papers is shown in Question 2 on p. 78.

Sex linkage

One of the genes that controls eye colour in fruit flies is on the X chromosome. Some fruit flies have white eyes because there has been a large deletion of the gene. Table 5 shows the five different genotypes and their phenotypes.

Table 5 Genotypes and phenotypes for eye colour in *Drosophila melanogaster*

Males		Females	
Genotype	**Phenotype**	**Genotype**	**Phenotype**
X^RY	Red eyes	X^RX^R	Red eyes
X^rY	White eyes	X^RX^r	Red eyes
		X^rX^r	White eyes

Note that the males are **hemizygous** — they only have one allele which is always expressed in the phenotype. When the mutant allele is recessive, heterozygous females are described as **carriers**.

The X and Y chromosomes are always included in genetic diagrams involving sex linkage — see Figure 11, which shows the inheritance pattern when a pure-bred (homozygous) white-eyed female is crossed with a red-eyed male. Notice:

- how eye colour 'switches' between the sexes in the F_1 generation
- the ratio in the F_2 generation

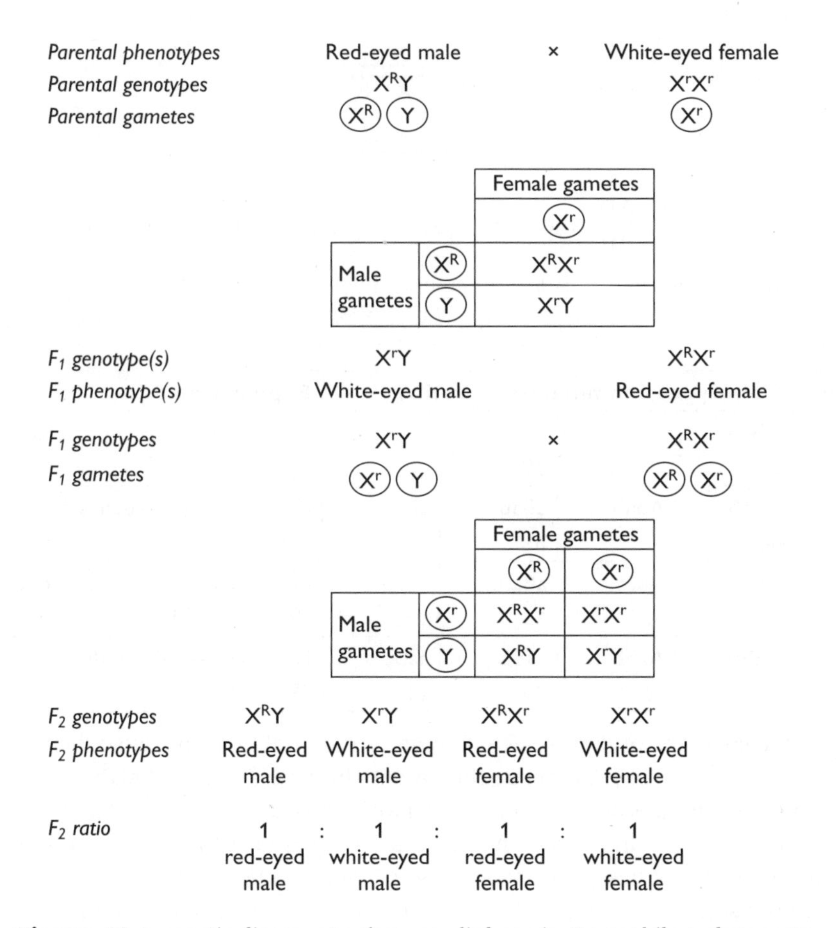

Figure 11 A genetic diagram to show sex linkage in *Drosophila melanogaster*

Features of sex linkage include:

- The expression of the sex-linked mutant phenotype is more common in males.
- Male offspring cannot inherit the trait from their fathers, but female offspring can.
- Female offspring may be carriers; male offspring can never be carriers.
- Males inherit their Y chromosome from their father. This chromosome does not have some loci that are found on the X chromosome.

Linkage

When the terms 'linkage' and 'linked' are used about genes without mention of 'sex', then this means the genes exist on the same autosome (non-sex chromosome). With approximately 14 000 genes and only five different chromosomes (one of which is tiny), genes in *Drosophila* must be linked. It is estimated that humans have up to 30 000 genes on 23 different chromosomes. Chromosome 1 (the largest) has over 4000 genes.

Linkage reduces variation because genes are inherited together in linkage groups. When homologous chromosomes pair during meiosis they form chiasmata between them and break and exchange portions of non-sister chromatids during **crossing over**. Table 6 shows the results of some test crosses with fruit flies.

Knowledge check 16

What results would you expect in the F_1 and F_2 if the parents were a white-eyed male and a homozygous red-eyed female?

Knowledge check 17

Haemophilia and colour blindness are X-linked recessive conditions. A grandfather has both conditions. Three of his grandsons have both conditions, but a fourth is colour blind only. No-one in the rest of the family has these conditions. Explain how this happened.

Examiner tip

To follow the sections on linkage and epistasis you must understand the pattern of inheritance for two genes as shown in Figures 9 and 10. Write out a genetic diagram, including a Punnett square, for the imaginary cross: **AABB** × **aabb** in the F_1 and F_2 generations.

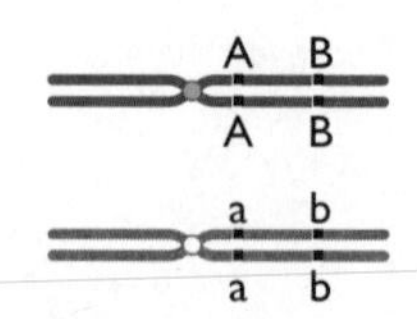

This is abbreviated to $\frac{\textbf{AB}}{\textbf{ab}}$, and in a genetic diagram, to **ABab**, rather than **AaBb**.

Table 6 Hypothetical results for test crosses involving female fruit flies heterozygous for two gene loci. In crosses 1, 2 and 3 the gene loci, A/a and B/b are linked on the same chromosome. In cross 4, the gene loci are on different chromosomes

Cross	Genotypes and phenotypic ratios				Explanation
1	ABab 52		abab 48		Complete linkage — no crossing over
2	ABab 46	Abab 5	aBab 6	abab 43	Partial linkage — genes close together
3	ABab 34	Abab 12	aBab 16	abab 38	Partial linkage — genes are further apart than in cross 2
4	AaBb 24	Aabb 26	aaBb 29	aabb 21	Unlinked — independent assortment

The genotypes in the pink boxes are known as parental types because they have the same genotype (and phenotype) as the parents in the test cross (**AaBb** × **aabb**). The genotypes in the blue boxes are recombinant types as they are different from the parents. In crosses 2 and 3, recombinants are the result of crossing over. In cross 4 they are the result of independent assortment (see Figure 10). If the gene loci are far apart on the same chromosome there will be nearly as many recombinants as might be expected with two unlinked genes.

Figure 12 shows how the genes are separated by crossing over.

Knowledge check 18

Test cross data give the following results: **AaBb** (49), **Aabb** (35), **aaBb** (40) and **aabb** (55). Do the data differ significantly from the expected 1:1:1:1 ratio?

Knowledge check 19

Crossing over does not occur in male *Drosophila*. What effect does this have on variation?

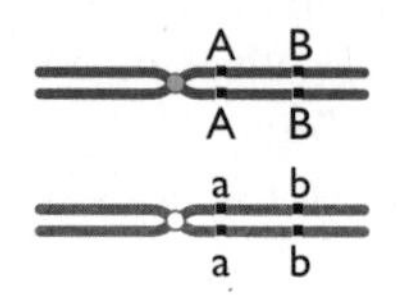

Pairing of bivalent in early prophase I

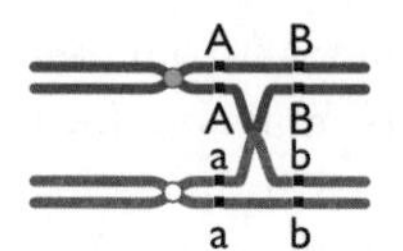

Chiasma forms between non-sister chromatids in prophase I

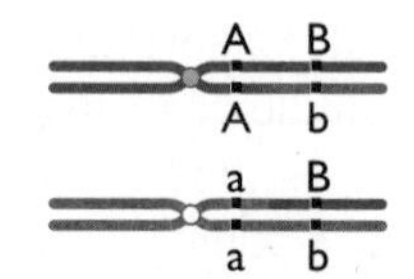

Breakage and exchange of parts of non-sister chromatids

Genotypes of gametes = (AB) (Ab) (aB) (ab)

Figure 12 Crossing over during prophase I

Epistasis

Epistasis concerns two (or more) genes that influence the same characteristic. Some genes code for enzymes. Most enzymes work as part of multi-step pathways, such as glycolysis. If an individual is homozygous recessive for an allele that codes for a non-functional enzyme then none of the enzymes that catalyse reactions later in the pathway can function because they have no substrate. When this happens it alters the expression of genes that code for enzymes further on in the pathway. Figure 13 shows four such interactions between genes in hypothetical pathways to produce flower pigments. The precursor compounds do not give any colour to the flowers, so they appear white.

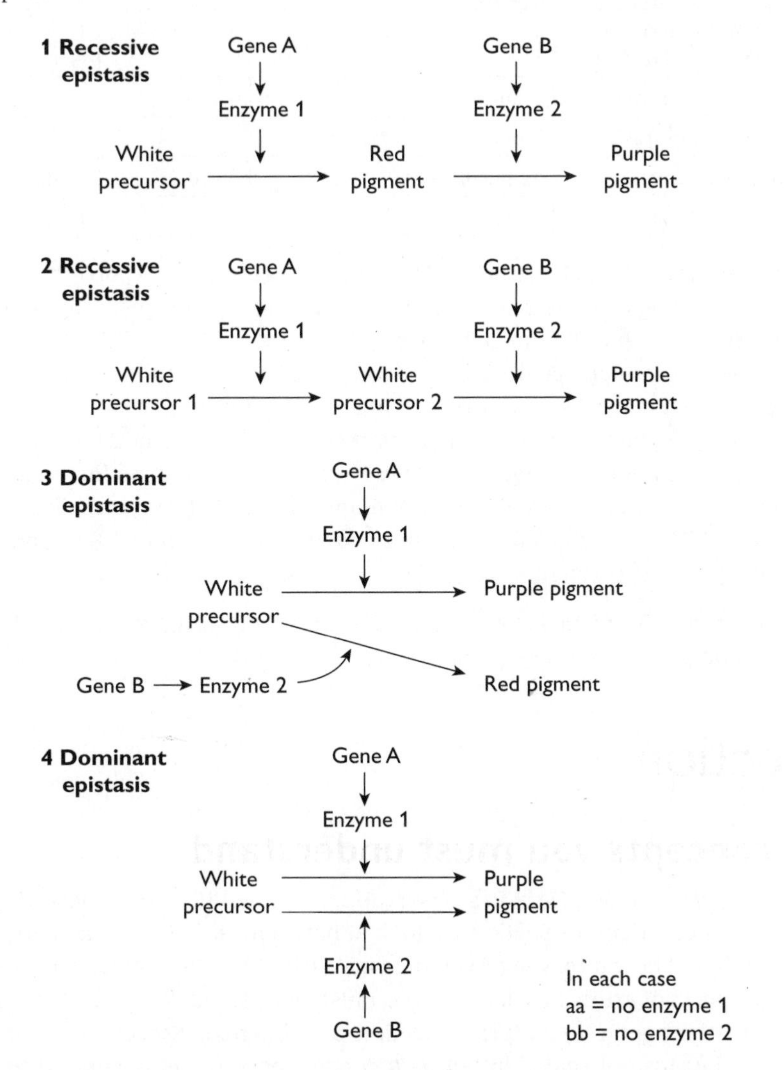

Figure 13 The action of two genes that code for enzymes in four different hypothetical pathways in production of flower pigments. In all cases the genes are not linked

Table 7 shows what happens to the variation in the F_2 generation in the four forms of epistasis shown in Figure 13. To understand this you need a Punnett square from the cross **AaBb** × **AaBb** suggested on p. 25. Epistasis reduces the variation that would be possible. Dominance also reduces variation; multiple alleles at one locus and codominance between alleles increase variation.

Table 7 The effect of epistasis on the phenotypic ratios in crosses between plants shown in examples 1 to 4 in Figure 13 when plants heterozygous for both loci (AaBb) are self-fertilised

Example		Ratios of phenotypes in F_2			
		A_B_	**A_bb**	**aaB_**	**aabb**
Genes do not interact, i.e. *not* epistasis (see Figure 9)		9	3	3	1
1	Recessive epistasis	9 purple	3 red	4 white	
2	Recessive epistasis	9 purple	7 white		
3	Dominant epistasis	12 purple		3 red	1 white
4	Dominant epistasis	15 purple			1 white

- **Example 1** Both **A** and **B** are needed to produce purple, but in **A_bb** red is produced as no functioning enzyme 2 is made. But when there are only recessive alleles (**aa**), neither **B** (to give purple) nor **b** (to give red) is expressed.
- **Example 2** Both **A** and **B** are needed to produce any pigment (purple). If one locus is homozygous recessive (**aa** or **bb**) then no pigment is produced.
- **Example 3** Enzyme 1 competes more successfully than enzyme 2 for the substrate. This has come about by gene duplication — there are two gene loci that code for variants of the same enzyme (there are many 'families' of enzymes like this).
- **Example 4** The two duplicate enzymes coded for by dominant alleles both act to produce the same pigment.

In some examples of epistasis, one of the gene loci produces an enzyme inhibitor (see Question 2 on p. 78).

Selection

Key concepts you must understand

In Unit F212, you studied Darwin's observations that led him to propose the theory of natural selection. In this section we look at how natural selection acts to stabilise a population and how it may act to change it. Much of Darwin's evidence came from studying aspects of artificial selection. You must understand that while the principles are similar, the agents of natural selection are different. Sometimes populations change not because of selection but as a consequence of being small (see genetic drift on p. 30).

Key facts you must know

Changes to allele frequencies

If an environment stays constant, then selection acts to maintain the population. Selection against extremes in features that show continuous variation maintains the same distribution from generation to generation. The range does not decrease with time because of mutation, meiosis and the influences of the environment. The example of **stabilising selection** in Figure 14 is from a study of birth mass in human babies published in 1973. This shows that mortality in very small and very large babies was high, so maintaining the variation in birth mass shown from generation to generation. Features where selection favours the heterozygotes also leads to stabilising selection; an example is the maintenance of the Hb^S allele in human populations in places where malaria is found (see Table 3 on p. 16).

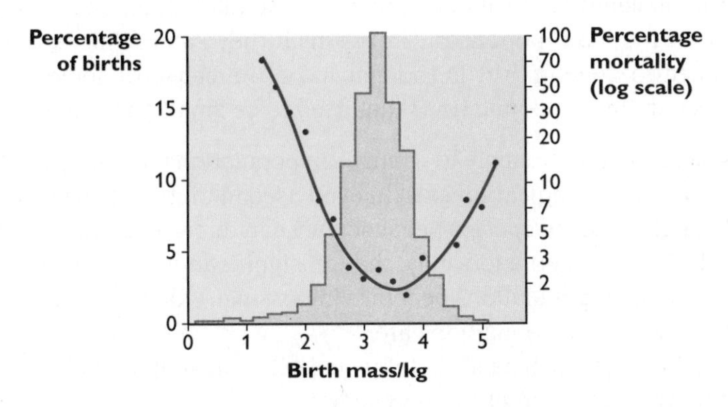

Figure 14 Stabilising selection in size of human babies

Changes in abiotic and biotic factors in the environment change the selective pressures that act on organisms. Certain individuals may show features that are adaptations to the changed environment. These individuals have a competitive edge and are able to survive, breed and pass on their alleles. This gives rise to **directional selection**. The medium ground finch *Geospiza fortis*, that lives on islands in the Galápagos, feeds on seeds. In the 1970s, there was a drought that killed many plants. Those that survived had large seeds. Many finches died, but the survivors had beaks that were 4% larger. The mean and range of the population shifted, as is shown in Figure 15. This change was associated with changes in genes for the development of the beak.

Examiner tip

Studies on the wing length of reed warblers and sparrows have shown that those with wing lengths at or near the mean length tend to survive longer and in larger numbers than those with long and short wings.

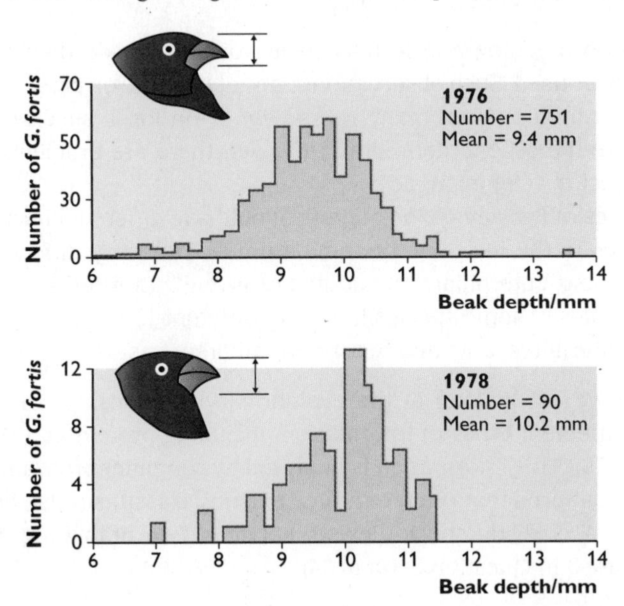

Figure 15 Directional selection in the ground finch, *Geospiza fortis*, on Daphne Major island in the Galápagos between 1976 and 1978

Genetic drift

Examiner tip
See p. 33 for an explanation of the concepts of the gene pool and allele frequency.

In small populations, changes in allele frequency can occur at random. It may be pure chance as to which individuals survive and breed. This change in allele frequency is known as **genetic drift**. Small populations that colonise new areas, especially islands or other isolated ecosystems, may also have allele frequencies that are not representative of the main population from which they came. This is the **founder effect** and allied to genetic drift is thought to be the cause of some of the allele frequencies seen in human populations that used to be small and isolated.

The changes described so far apply to changes in populations of a species. According to the definition of the biological species that you used in Unit F212, these populations are all members of the same species because they can interbreed and produce viable offspring. With time, a population may change significantly in other ways so that reproduction with other populations becomes impossible. Examples of these isolating mechanisms that separate populations are:

- geographical barriers, such as mountains and seas, prevent individuals mating
- breeding occurring at different times of year
- courtship rituals being different
- sterile hybrids being unable to breed

Species definitions

If species can change over time, this begs the question 'what is a species'? In Unit F212 you learnt the definition which is known as the **biospecies** — a group of organisms that are able to interbreed and give rise to fertile offspring.

Another way of looking at species is the **phylogenetic species** — a group of organisms sharing many characteristics that are the basis for its classification and reflect its phylogeny (evolutionary history).

Examiner tip
You should read about recent discoveries of living and extinct organisms new to science to understand how difficult it can be to name and classify them.

If observations on breeding and fertility have not been made then the biospecies definition cannot be used. Such observations cannot be made on preserved specimens collected on expeditions (often many years ago) or on fossil species. Taxonomists often have to use the second definition. However, there are problems in assigning organisms to species — for example:

- different species of *Drosophila* look identical but have different courtship rituals
- different stages in the life cycle of some animals look very different and do not breed together, e.g. caterpillars and adult moths and butterflies
- males and females of some species look very different
- social insects (termites, ants and bees) have different castes

Cladistics is a way of looking at the evolutionary relationships between species that uses many different types of information, including protein and DNA sequence data (see pp. 44–45). This information is analysed by computer programs. The results show branching patterns that often reinforce existing classifications, but sometimes challenge them. A cladistic species exists between two branching points in this system (see Figure 3 in Question 3, on p. 84).

Artificial selection

In natural selection the agents of selection are biotic and abiotic aspects of the environment. In artificial selection humans are the agents of selection because they choose which individuals will survive and breed to pass on their alleles. They often also decide the matings that will occur.

Knowledge check 20

Explain why it says '... breed to pass on their alleles' rather than '... breed to pass on their genes'.

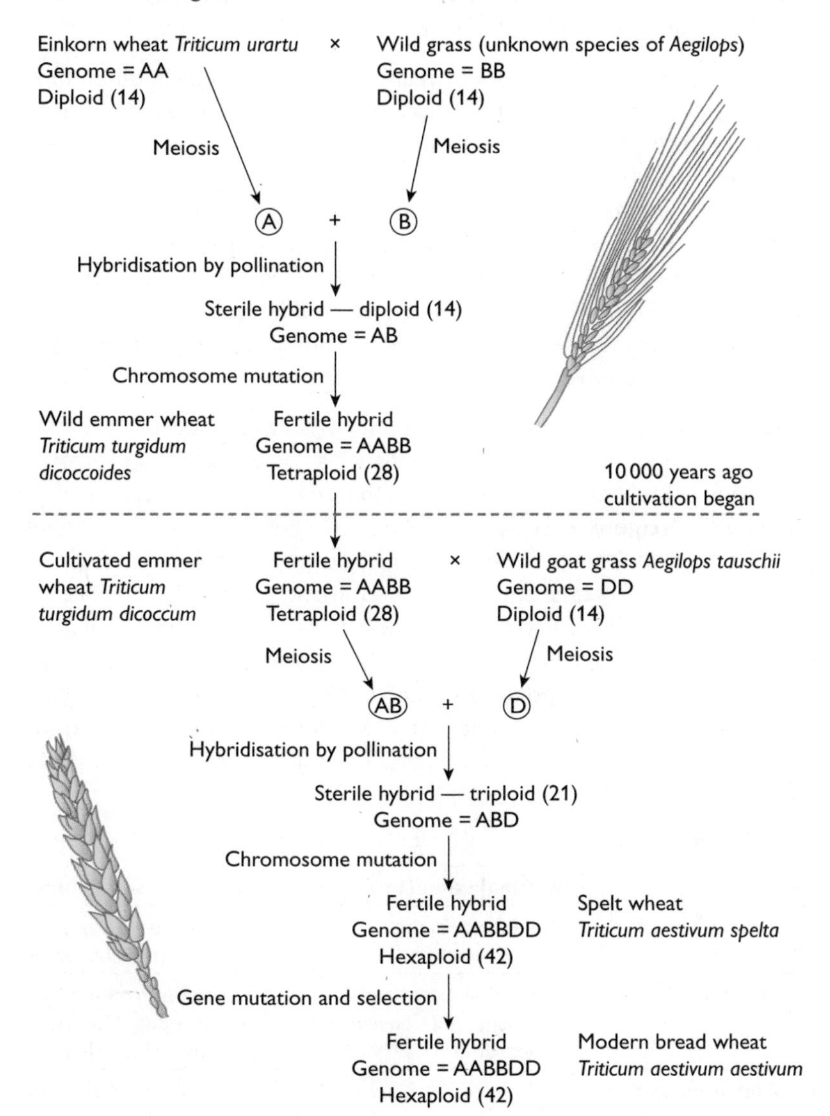

Figure 16 Evolution of modern bread wheat by hybridisation and polyploidy. Each genome (A, B and D) has seven different chromosomes

Knowledge check 21

State the number of chromosomes in the gametes of each of the species shown in Figure 16.

Artificial selection has been involved in the evolution of modern bread wheat, *Triticum aestivum* (see Figure 16). This has involved:

- farmers in the Middle East 10 000 years ago collecting grain from wild emmer wheat and choosing certain features, such as non-brittleness of the ears, which changed the frequencies of many alleles in the gene pool and gave rise to cultivated emmer wheat

- hybridisation between various species, which occurred at least twice and produced sterile hybrids
- chromosome mutation, which doubled the chromosome number so plants became polyploid and fertile as chromosomes could pair in meiosis
- the introduction of genomes (known as A, B and D) from different species, which introduced new genes — for example, the genes for glutens (proteins needed for bread making) were in the genome of *Aegilops tauschii*

Plant breeders improved bread wheat by incorporating genes from different species. The gene ***Sr2***, responsible for resistance to stem rust, a fungal disease, was transferred from *T. turgidum* into common (hexaploid) wheat to produce a variety called Hope. It has since been used in other varieties. Mutant alleles that do not code for the enzymes required to make gibberellins have also been incorporated to give dwarf varieties that have increased grain yield. In this process of genetic improvement, the high yielding variety is crossed with a variety showing the desired characteristic. The progeny are tested to see if they have the desired feature and then crossed with the high yielding variety. This backcrossing occurs for several generations before the new variety is made available to farmers.

Examiner tip
Gibberellins are plant growth substances. There is more about them on pp. 61–62.

Selection is different with domesticated animals (e.g. cattle) because they are much larger and have a longer life cycle than a plant such as wheat. Milk yield is an example of continuous variation that makes identifying the success of the breeding programme more problematic. **Progeny testing** is used to find out if a male possesses alleles that will improve milk yield. This is done by seeing how good his female offspring are at producing milk, which will take several years. However, sperm can be collected from potentially good males and stored for later use — possibly long after the male is dead. The sperm is stored and delivered to suitable females by artificial insemination (AI). Finding suitable females uses **performance testing** to record their milk yields. The animals are kept under the same conditions to reduce the environmental influence on variation. Eggs from high-performing females may be collected, fertilised in vitro and inserted into surrogates (see p. 37).

Table 8 Natural selection and artificial selection (selective breeding)

Feature	Natural selection	Artificial selection
Agent of selection	The environment	Humans, e.g. farmer, grower, plant or animal breeder
Features selected	Those that adapt the organism to its environment at that time	Those determined by the breeder that are commercially valuable
Number of features involved	All	A small number, e.g. milk yield
Allele frequencies in the gene pool	Change as a result of selection for or against phenotypes	Change as a result of selection and non-random mating determined by humans
Speed	Slower	Faster

Artificial selection is not restricted to plants and animals. Research scientists use similar methods to improve strains of bacteria and yeasts for industrial processes, such as yoghurt production, brewing and baking.

Focus on mathematical skills: the Hardy–Weinberg principle

Allele frequency is used as a measure of the degree of selection that occurs. To understand this you need to appreciate the concept of the **gene pool**. This is all the alleles of a particular gene that exist in a population. Each individual is diploid, so may contribute two alleles (for each gene) to the gene pool.

The Hardy–Weinberg principle is used to calculate the frequencies of alleles. Equation 1 is for alleles; equation 2 is for genotypes:

Equation 1

$$p + q = 1$$

p = the frequency of the dominant allele, **A** (or whatever allele is being investigated), in the population

q = the frequency of the recessive allele, **a**, in the population

You can use a modified Punnett square to show how equation 2 is derived. Remember that p and q represent the frequencies of alleles in the gene pool. So this shows the likely outcomes if all individuals can mate at random.

		Frequency of female gametes in the gene pool	
		p (A)	**q (a)**
Frequency of male gametes in the gene pool	**p (A)**	**p^2 (AA)**	**pq (Aa)**
	q (a)	**pq (Aa)**	**q^2 (aa)**

Equation 2

$$p^2 + 2pq + q^2 = 1$$

p^2 = the frequency of the genotype **AA**

$2pq$ = the frequency of the genotype **Aa**

q^2 = the frequency of the genotype **aa**

(If there are three alleles of the gene then the equations are expanded, thus:

$$p + q + r = 1$$

$$p^2 + 2pq + 2pr + 2qr + q^2 + r^2 = 1)$$

If we investigate a population of mice and there are 16% with the recessive phenotype (they have the genotype **aa**) we can say that the frequency of the recessive allele **a** is the square root of 0.16, which is 0.4. This means that the frequency of the dominant allele **A** is 0.6 ($p = 1 - q$). We can then calculate the frequency of the homozygous dominant genotype as **AA** = $0.6^2 = 0.36$ (36%) and of the heterozygous genotype as **Aa** = $2 \times (0.6 \times 0.4) = 0.48$ (48%).

We can do this because the homozygous recessive individuals in the population have a recognisable phenotype. We cannot count the number of homozygous dominant individuals to find p^2 because their phenotype is indistinguishable from

Knowledge check 22

Cystic fibrosis is a recessive condition that affects about 1 in 2500 babies in the UK. Use this information to calculate: (a) the frequencies of the dominant and recessive alleles, **F** and **f**, in the UK population, and (b) the percentage of heterozygous individuals (carriers) in the population.

Knowledge check 23

The frequency of the blood group alleles in a population are: $\mathbf{I^A} = 0.3$, $\mathbf{I^B} = 0.1$, $\mathbf{I^O} = 0.6$. Calculate the frequencies of the three blood groups in the population.

that of the heterozygotes. We can use the equations to calculate the frequency of **A** in the population and therefore the frequencies of **AA** and **Aa**.

The Hardy–Weinberg principle assumes that:

- the population is large
- mating occurs within a population at random (mating in a small population is non-random)
- there is no mutation
- there is no immigration or emigration
- there is no selective pressure operating against one of the phenotypes

The Hardy–Weinberg equations can be used to see if allele frequencies have changed. If the frequencies of genotypes do not conform to the Hardy–Weinberg principle then one or more of the conditions above does not apply. It may be that selection is occurring.

If you do an image search for Hardy–Weinberg equilibrium, you will find graphs that show the effect of changing allele frequencies on genotype frequencies. You can learn more about this and test your knowledge at:

www.phschool.com/science/biology_place/labbench/lab8/intro.html

Synoptic links

Selection was covered in Unit F212. It would be a good idea to revise all the aspects of this and the evidence for evolution (see pp. 60–65 in the guide for Unit F212). In order to understand the next section on cloning it is important to revise mitosis from Unit F211 (see pp. 22–23 in the guide for Unit F211).

Summary

- There are two stages in meiosis. During meiosis I homologous chromosomes pair, exchange parts of non-sister chromatids in crossing over and then arrange themselves on the equatorial plate. The nuclear membrane breaks down and centrioles assemble microtubules into the spindle. During anaphase I, homologous pairs separate to halve the chromosome number in the nuclei that form in telophase I. In meiosis II, chromatids separate in a division that is similar to mitosis.
- A gene is a length of DNA that codes for a polypeptide. Its locus is its position on a specific chromosome. Alleles are versions of a gene, but with different sequences of bases.
- Genotype refers to alleles of one or more genes. Phenotype refers to any feature of an organism apart from its genotype. Genotype and environment interact to influence the phenotype. Dominant alleles are always expressed in the phenotype; recessive alleles are only expressed when homozygous (e.g. **aa**). If two different alleles of one gene are both expressed in the phenotype, they are described as codominant.
- Linkage is the existence of two or more genes that have their loci on the same chromosome. Recombination between linked genes occurs as a result of crossing over. Genes that have their loci on the X or Y chromosome are sex linked.
- Causes of variation are: crossing over, independent assortment of unlinked allelic pairs in metaphase I of meiosis and the combinations of alleles at fertilisation.
- Epistasis refers to the interaction between gene loci. There are dominant and recessive forms of epistasis that influence phenotypic ratios in the offspring of genetic crosses.
- The chi-squared (χ^2) test is used to test the significance of the difference between observed and expected results in genetic crosses.l

- Continuous variation is variation in quantitative features, such as length and mass, whereas discontinuous variation is variation in qualitative features, such as colour and shape. Continuous variation is influenced by many genes (polygeny) and the environment; discontinuous variation is influenced by one or a few genes.
- Variation is the raw material for selection — both natural and artificial. Selective agents, such as predators, food supply and climate, can act to stabilise a population so it remains constant from generation to generation or change it so it is adapted to changes in the environment.
- The Hardy–Weinberg principle is used to calculate the frequencies of alleles in populations. These frequencies stay the same from generation to generation if there is no selection, mutation, migration or non-random mating.
- Large changes in allele frequency can occur by genetic drift, which is the result of non-random mating within small populations.
- The biological species concept is difficult to apply to many species as mating and the production of fertile offspring are not easily observed. The phylogenetic species concept relies on observable features and is commonly used.
- Ecological (geographic), seasonal (temporal) and reproductive mechanisms act to isolate populations of different species.
- Humans choose features to improve in domesticated animals and plants, such as dairy cattle and bread wheat. They choose the individuals that will breed and then select from among the offspring those that will be used to breed the next generation.

Summary continued

Biotechnology and gene technologies

Cloning

Key concepts you must understand

If a plant or animal showing all the desirable features is allowed to outbreed, then the offspring will not be identical to the parent. This is because of the causes of variation that are listed on p. 17. If the organism is self-fertilised, then the offspring will be more like the parent. This is possible with self-pollinating plants, such as wheat. With animals, the best that can be done is to cross an individual showing the desired features with another that is similar. If this continues for generation after generation it leads to inbreeding. Our cereal crops (wheat, barley, maize) are much inbred so that different varieties only have a few gene loci that are different. Genetic faults in animals, such as hip dysplasia in some breeds of dogs, are often the result of inbreeding. Dogs with hip dysplasia are lame in their back legs.

Examiner tip
Outbreeding is the breeding of individuals that are not closely related.

Some organisms reproduce asexually so it is possible to perpetuate a strain that shows desirable features. For example, strawberry plants produce runners, which take root and separate from the parent plants. Many crop plants, including all cereals, do not do this and few animals reproduce asexually. The answer is cloning. **Reproductive cloning** produces *new individuals* with the genotype of the existing individual. Do not confuse this with **non-reproductive cloning** which is carried out to make cells to replace those of an individual — the genotype of the cloned cells matches that of the existing individual.

Key facts you must know

Cloning plants

English elm trees, *Ulmus procera*, are all genetically identical. The Romans brought a single clone of elm trees to Britain for the purpose of supporting and training vines. *U. procera* reproduces by growing suckers from the roots — although it flowers and produces fruits, no seeds are produced. Therefore, apart from some mutation, asexual reproduction has preserved this clone unchanged for 2000 years. The suckers can be cut off, planted out and grown on. Any form of asexual reproduction involving growth of new parts that can separate from the parent is known as **vegetative reproduction**. When used by humans it is known as **vegetative propagation**.

Tissue culture is a way of producing plants that do not reproduce naturally by asexual means or are hybrids that do not breed true. It is a technique that is useful for ornamental and rare plants, such as orchids. Parts of a plant showing desirable features are removed and used to produce more individuals, as shown in Figure 17.

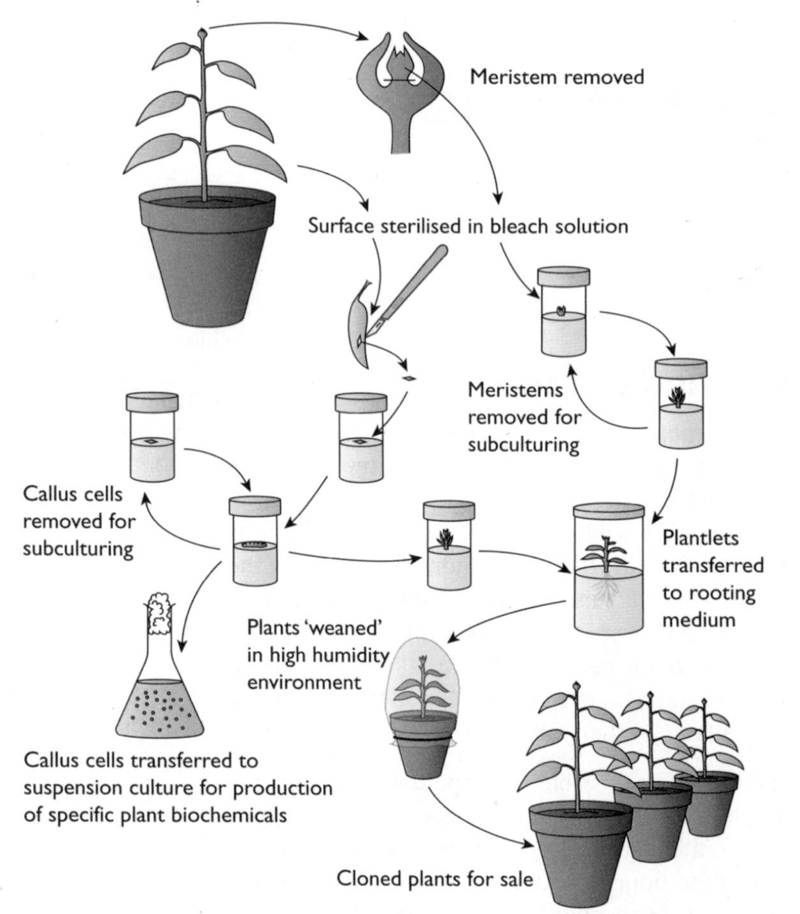

Meristem culture
Meristems contain unspecialised cells that divide by mitosis. Viruses tend not to infect meristems, so meristems can be removed from plants and used to produce virus-free material for cloning. Meristems grow into plantlets that have buds with more meristems that are subdivided to repeat the process.

Callus culture
A callus is a mass of undifferentiated tissue that forms when pieces of shoot, root or leaf are placed in tissue culture. Callus cells can be maintained in tissue culture and subdivided to give large quantities of tissue for cloning. If the growth medium contains appropriate concentrations of plant growth substances (auxins and cytokinins), roots or shoots will develop to give genetically identical plantlets.

Suspension culture
Plant cells from callus cultures are separated from one another and maintained in suspension in a liquid medium. The red dye shikonin is made by plant cells cultured in bioreactors.

Figure 17 Plant tissue culture

Some advantages and disadvantages of cloning plants are shown in Table 9.

Table 9 Advantages and disadvantages of cloning plants in agriculture and horticulture

Advantages	Disadvantages
• Uniform plants make harvesting easier; uniform qualities of crops • Makes clones of plants that cannot reproduce sexually to set seed, e.g. hybrids of lavender, banana plants • Can be used to clone transgenic plants • Allows stocks to be built up quickly; does not have to rely on slower sexual methods of propagation • Tissue culture can be set up anywhere and small plantlets can be transported easily	• All are susceptible to the same pathogen or pest species, or to changes in climate, etc. • Propagates single clones that may have genetic diseases and/ or do not have resistance to some diseases • Conditions in tissue culture must be kept sterile — aseptic techniques must be used and infected cultures thrown away • Propagating plants vegetatively is labour-intensive and tissue culture requires trained staff and expensive facilities

Cloning animals

Dairy cattle are cloned by using hormones that stimulate the ovaries to produce a large number of eggs. This is known as **superovulation**. These eggs are harvested from the ovary and fertilised in vitro by sperm from a superior bull. The zygotes divide by mitosis and the resulting embryos are sexed (to make sure they are XX), subdivided several times and implanted into surrogate mother cows. This means that all the resulting calves are clones of *each other* but not of their mother. To clone a high performing animal this method must be modified as shown in Figure 18, which was the method used to produce Dolly the sheep. Some advantages and disadvantages of this method are shown in Table 10.

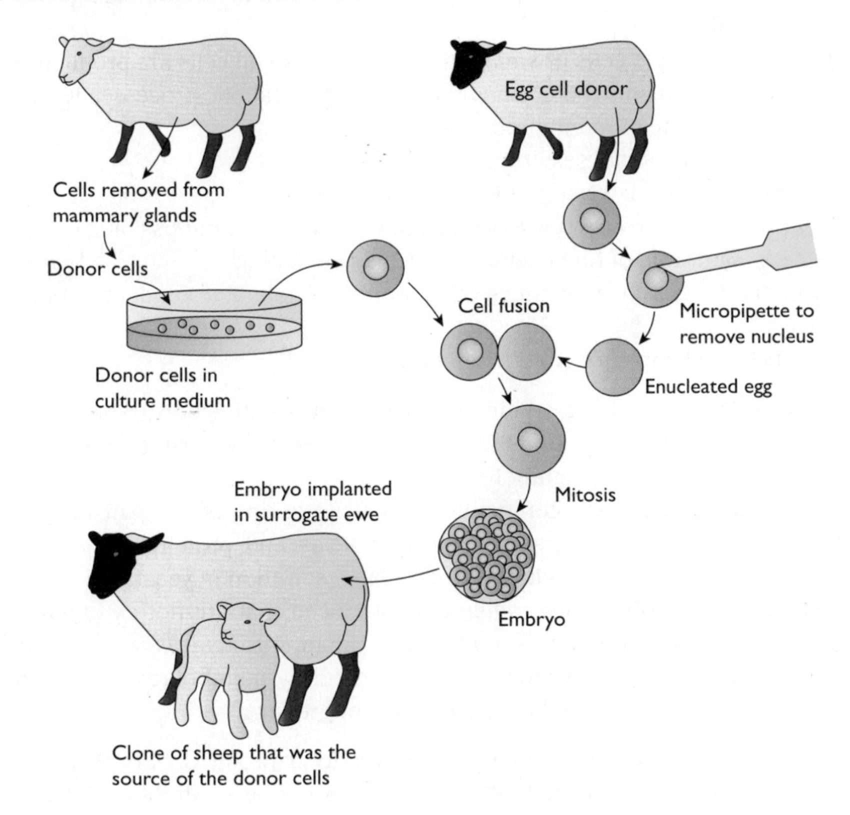

Figure 18 The steps involved in cloning sheep and cattle

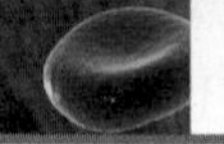

Table 10 Advantages and disadvantages of cloning animals

Advantages	Disadvantages
• Animals giving high yields can be produced, e.g. for high milk yield or quality of meat • Prevents rare animals becoming extinct with loss of their genes/alleles • Fast reproduction of transgenic animals, e.g. sheep that produce human proteins in their milk	• Animals may have low quality of life • Animals become genetically uniform, which increases susceptibility to disease • Success rate has been very low

Therapeutic cloning may prove to be a successful way to treat diseases, such as heart disease and diabetes. Stem cells divide to produce specialised cells to replace damaged tissue. This is a non-reproductive form of cloning.

Biotechnology

Key concepts you must understand

Biotechnology is the use of living organisms in the production of useful products, such as foods and drugs, or in useful services, such as sewage disposal. Biotechnological processes use either whole microorganisms — bacteria, archaeans, yeasts, mould fungi and protoctists — or enzymes.

Microorganisms are used in these industries for the following reasons:

- They grow rapidly (see Figure 19) and reproduce asexually to produce genetically identical populations.
- They remain as single cells or small groups of cells, so all cells are productive.
- Among microorganisms there is a wide variety of metabolic processes with much potential for making different products.
- Strain selection (artificial selection) is easy.
- Genetic modification is easy (see Figure 23 on p. 46).
- They carry out complex processes that would be difficult or impossible by chemical means, e.g. synthesis of fine chemicals such as penicillin.
- It is possible to use cheap sources of nutrients, often waste products from other industries.
- They do not need complex growth conditions.

In some cases the product is a whole organism, as it is with Quorn™ (mycoprotein) production that you studied in Unit F212. In most cases the product is a metabolite — a chemical produced during metabolism.

- **Primary metabolites** are compounds produced during the organism's normal metabolism for its survival, such as amino acids and proteins. Ethanol is an example — it is the waste product of anaerobic respiration in yeast.
- **Secondary metabolites** are compounds produced that are not essential to the organism's survival. Penicillin is produced by the fungus *Penicillium chrysogenum* when it stops growing. The production of secondary metabolites may provide a way of using enzymes while the organism is not growing.

One disadvantage of using microorganisms for food production is that in most biotechnology processes only one organism or a particular strain of an organism is required. The culture must not become colonised and contaminated by competitors or parasites, such as viruses, which may kill the organism. Contaminants would use up the nutrients in the culture and might produce substances that are distasteful or toxic.

Key facts you must know

A starter culture of bacteria is put into a sterile culture medium in a sterile container. Aseptic technique is used throughout to avoid the culture becoming contaminated by competitors and parasites. The culture medium and glassware used are sterilised by placing them in an autoclave that applies steam under pressure. The top of the sample container is passed through a Bunsen flame that creates an updraft of air so that bacteria from the air do not enter the culture. Samples of the culture are taken at intervals and the number of bacteria counted. The samples are used to estimate the population in numbers per mm^3. Figure 19 shows the results.

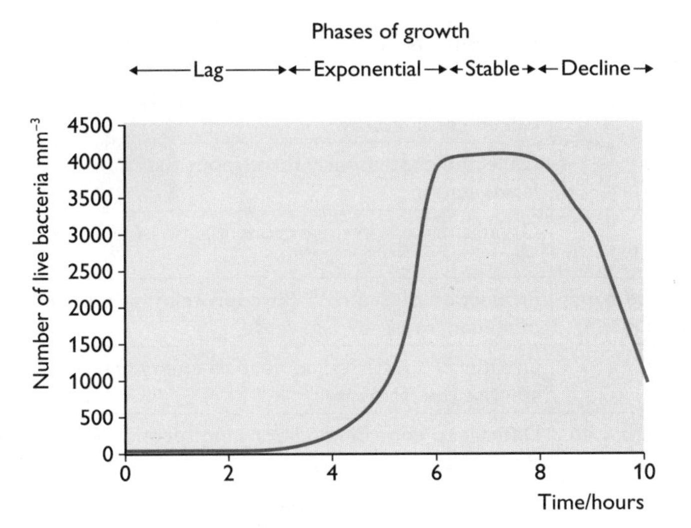

- **Lag phase** Growth is slow or non-existent. Reproduction rate is slow as cells are absorbing nutrients, producing enzymes and storing energy.
- **Exponential (log) phase** Reproduction rate is fast. Few cells die. Some bacteria reproduce once every 20–30 minutes so the population doubles in this amount of time. To fit the numbers onto a graph it is usual to plot the $\log_{10}$ (log to the base 10) of the numbers against time (log × linear graph).
- **Stationary (stable) phase** A decelerating phase leads into a period when the population remains constant because death rate equals reproduction rate.
- **Decline phase** The death rate is greater than the reproduction rate. The population may decline to zero.

Figure 19 The changes in numbers of bacteria in pure culture. The lag phase to stationary (stable) phase is called sigmoid growth

The pattern described in Figure 19 can be explained in terms of the factors that limit growth, which are:

- availability of nutrients
- availability of oxygen
- presence of waste products, some of which may be toxic (e.g. ethanol)

During the lag and exponential phases there are plenty of nutrients and dissolved oxygen. These decrease as the population increases and limit population growth. The accumulation of waste, such as carbon dioxide and ethanol, may make the environment unsuitable so cells stop reproducing or die.

To find the population of live bacteria, samples are incubated on agar plates and the number of colonies is counted. Each colony represents one bacterial cell in the original sample. Alternatively, the sample may be treated with a dye that stains live cells which can then be counted under a microscope.

Knowledge check 24

An imaginary bacterial culture starts with one bacterium. The rate of division is once every 20 minutes. Calculate the numbers of bacteria at every 20-minute interval over 5 hours.

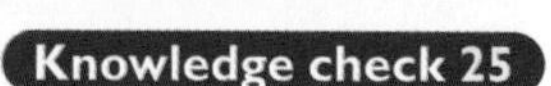

Knowledge check 25

Draw a graph of your results from Knowledge check 24. To fit the numbers onto a graph plot the log_{10} (log to the base 10) of the numbers against time (log × linear graph). Search online for this graph paper, print it out and plot the points. What sort of line do you get?

In a **batch culture**, a starter culture of yeast or bacteria is added to an appropriate substrate and the fermentation runs for the length of time needed for the organism to complete the sigmoid growth pattern. At the end of that time the fermenter is emptied, sterilised and the whole process started again with a new batch. **Continuous culture** is used in the manufacture of Quorn™. It involves maintaining a culture of the filamentous fungus *Fusarium venenatum* in the exponential phase by supplying nutrients continuously. Fermentation broth containing fungal hyphae (the product) is removed continuously. Batch and continuous culture are compared in Table 11.

Conditions inside fermenters are maintained or even changed during the fermentation to give the best growing conditions for the microorganism. This maximises the yield.

Products may undergo filtration, purification and packaging, collectively called **downstream processing**.

Table 11 Comparison between batch culture and continuous culture

Feature	Batch culture	Continuous culture
Fermenter	Closed	Open
Nutrients	Added at the start of fermentation	Added continuously
Product	Collected at the end of fermentation	Collected continuously throughout the fermentation
Growth phase of microorganism	Exponential phase is short	Organisms kept in the exponential phase
Examples of products	Wine, beer, yoghurt, penicillin (in a fed-batch culture), enzymes for washing powders	Production of Quorn™ (mycoprotein); production of human insulin by GM yeast
Advantages	Easy to control conditions; produces secondary metabolites	Greater productivity; no need to empty and sterilise the fermenter
Disadvantages	Large vessels needed; waste builds up	Difficult to control conditions; no secondary metabolites produced

Knowledge check 26

Explain why the making of yoghurt, beer, bread and cheese are biotechnology industries.

In some biotechnology processes enzymes are used rather than whole organisms. This is when a one-step process is required, such as the production of fructose from glucose by the enzyme glucose isomerase. Many enzymes used in large-scale commercial processes are immobilised by being bonded to an insoluble matrix, held inside a gel lattice such as silica gel, held in tiny capsules of alginate or held inside a partially permeable membrane. Immobilised enzymes are more stable than non-immobilised enzymes when exposed to changing conditions of pH or temperature. They are removed easily at the end of fermentation so do not contaminate the product. This reduces the cost of downstream processing. They can also be reused many times.

Synoptic links

Fermentation describes the culture of microorganisms in bioreactors to produce useful products. The organisms may respire aerobically or anaerobically. Questions on biotechnology may require an understanding of respiration from Unit F214. Biotechnology processes provide many foods and food ingredients, e.g. monosodium glutamate. Questions may be set in the context of world food supplies from Unit F212.

Summary

- Reproductive cloning is the production of new individuals genetically identical to the parent organism. Non-reproductive cloning involves producing cells and tissues that are genetically identical to the parent organism, without forming new individuals.
- Plants that are propagated vegetatively form clones; elm trees are an example.
- Clones of plants are produced in tissue culture by taking explants from leaves, stems, roots or meristems. These are genetically identical so give uniform crops. All individuals of a clone may be susceptible to the same strain of a plant pathogen.
- Animal cloning is done by transplanting nuclei from an individual with required features to enucleated eggs, which are placed inside surrogate females. The animals produced show the desired features, but there may be problems with deformities and other health problems.
- Biotechnology is the use of living organisms (especially microorganisms), or cells taken from living organisms, to make products, such as foods and drugs.
- Microorganisms have simple requirements, can be kept in large containers, do not need large spaces for cultivation and can be genetically modified more easily than animals or plants. Aseptic techniques are used to avoid contamination by competitors and parasites.
- The growth of a microorganism in a closed (batch) culture is sigmoidal with lag, exponential, stationary and decline phases. Limiting factors, such as space, nutrients, temperature and pH, influence the growth and are maintained at optimum values in fermenters to maximise yields.
- Isolated enzymes are immobilised by encapsulating them in alginate beads or attaching them to surfaces. This allows enzymes to be reused in large-scale production processes, prevents them contaminating products and makes them more heat stable.
- Primary metabolites (e.g. amino acids, ethanol) are produced during growth of an organism; secondary metabolites (e.g. penicillin) are not required to support growth.
- At the end of a batch culture, a fermenter is drained, sterilised and refilled. The desired product is separated from the contents of the fermenter by filtration or centrifugation as part of downstream processing. Mycoprotein (Quorn™) is produced by continuous culture in which nutrients are continuously added and product (fungal hyphae) continuously removed. Organisms in continuous culture are held in the exponential phase.

Genomes and gene technologies

Key concepts you must understand

The **genome** refers to all the genetic material in an individual or in a species. Remember that all members of the same species have the same genes. Individuals differ genetically in the alleles that they have. Various discoveries in the latter half of the twentieth century made it possible to sequence DNA, to search for specific genes, to cut DNA, amplify it and transfer the products between genomes of different species. The discoveries gave rise to the following:

- **gene sequencing**, using knowledge of DNA replication
- **polymerase chain reaction**, using knowledge of DNA replication and an enzyme from an archaean that lives in hot springs
- cutting DNA by **restriction endonucleases**, which cut across both strands of DNA at specific sequences of base pairs (restriction enzymes protect bacteria by cutting the DNA of viruses that infect them)
- transfer of genes using **vectors**, such as viruses, plasmids and liposomes

Examiner tip

Liposomes are artificial vesicles made of phospholipids. They fuse with the cell membranes of target cells to deliver their contents. This is another opportunity for a synoptic question on cell membranes and biological molecules.

- producing DNA from an RNA template by **reverse transcription** using reverse transcriptase enzymes from retroviruses
- separating fragments of DNA by **gel electrophoresis**

Microbes, such as bacteria and yeasts, take up plasmids from their environment. This natural process gives them the opportunity to absorb new genes or new alleles. Plasmids often have genes for antibiotic resistance, so when bacteria exchange plasmids by conjugation susceptible strains can gain antibiotic resistance. **Genetic engineering** is the term usually applied to the process of genetic modification by means that are not possible using traditional breeding and artificial selection.

Key facts you must know

Proteins and fragments of DNA are separated by electrophoresis (Figure 20). This involves putting samples into wells cut into an agarose gel, which is in a tank filled with a buffer solution of an appropriate pH. A direct electric current is applied to the gel and the protein molecules or pieces of DNA move towards an electrode. DNA is negatively charged so moves towards the anode. The distance moved by a fragment depends on size; smaller fragments move further per unit time than larger fragments.

Melt agarose gel in buffer solution

Insert a toothed comb at one end of the tank to make the wells to take DNA samples

Pour in molten agarose gel

Leave gel to set. Place electrodes at either end of the tank

When gel is set pour in buffer solution and remove the comb

Add blue dye to each DNA sample

Add DNA and dye mixture to the wells

Connect electrodes to the power supply

When the blue dye is within 10 mm of the end of the gel, disconnect the power supply

Pour away the buffer and add DNA stain (Azure A) for 4 minutes

Rinse with water and analyse the fragments of the DNA which will appear blue

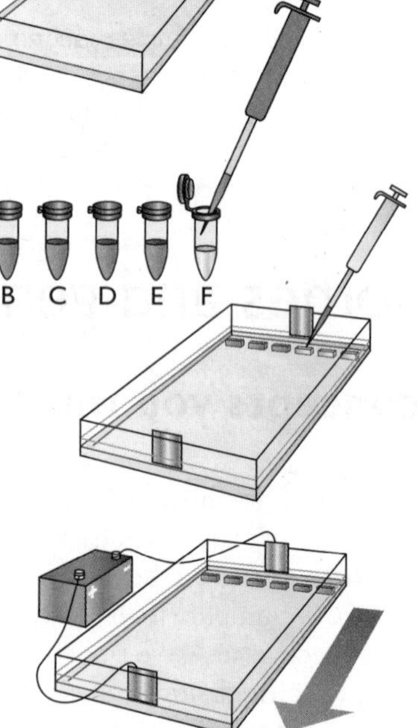

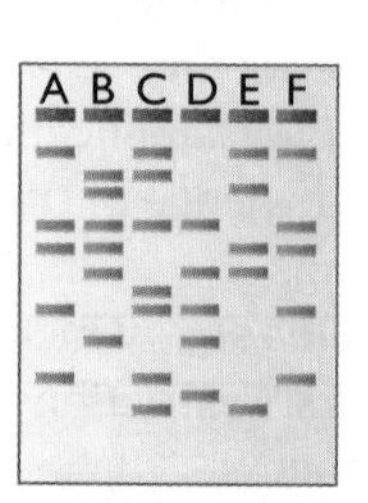

Figure 20 Electrophoresis

Examiner tip
If you do not get the chance to carry out an investigation using electrophoresis, watch a video or an online animation that shows how this is done.

Electrophoresis is similar to chromatography. If you have carried out chromatography on chloroplast pigments you will have seen the coloured pigments separating. DNA is invisible unless a blue or fluorescent dye is added. A radioactive DNA probe (with the isotope ^{32}P) may be used to locate specific sequences. These probes bind to complementary sequences in the DNA, making them show up as dark bands when exposed to X-ray film.

Samples of DNA for testing can be very small, for example those retrieved from crime scenes or from chloroplasts and mitochondria. Figure 21 shows how small samples of DNA are amplified in the polymerase chain reaction. Primers are short sequences of a polynucleotide that bind to the single-stranded DNA that is being copied. This is necessary for DNA polymerase to start the process of replicating the existing polynucleotide. *Taq* polymerase is an enzyme extracted from the thermophilic archaean *Thermus aquaticus* that lives in hot springs.

Examiner tip

The Archaea is a domain of organisms that you learnt about in Unit F212. Extremophiles, such as *T. aquaticus*, are classified in this domain. See p. 58 of the Unit Guide for F212.

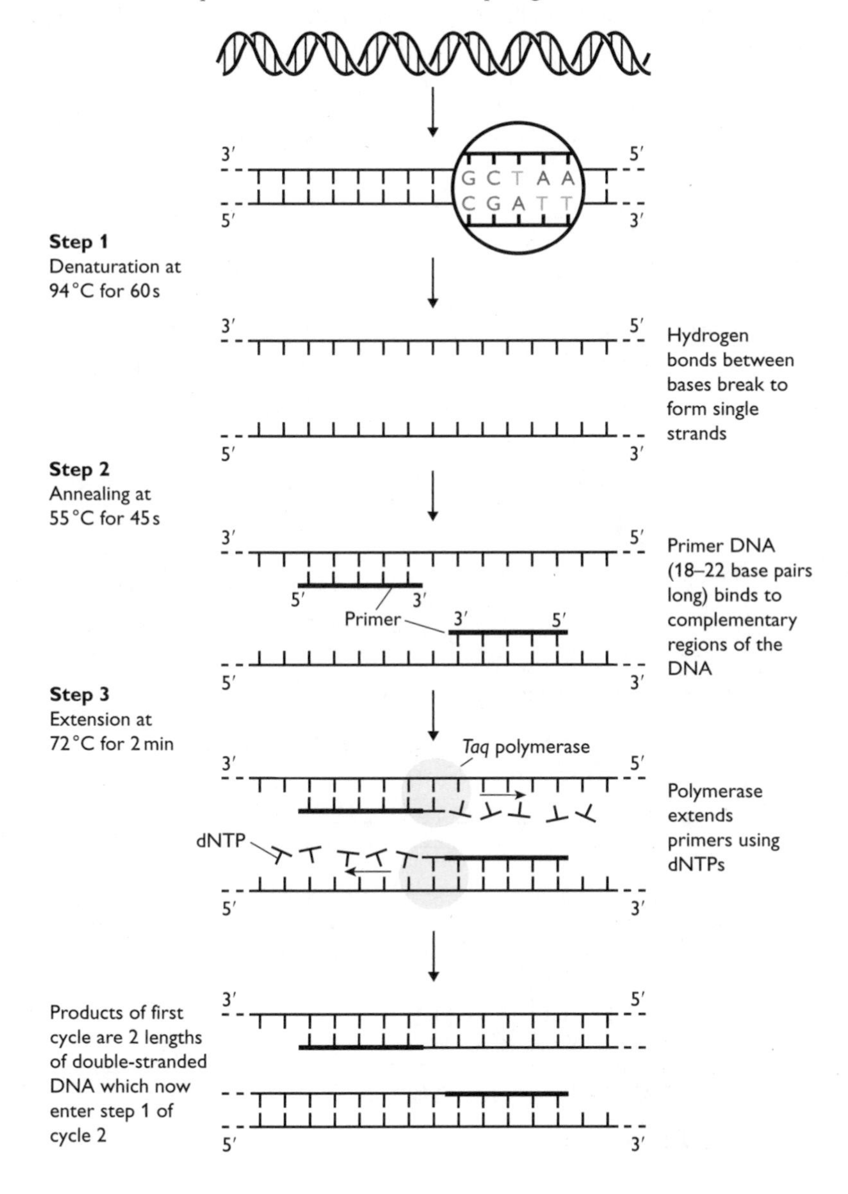

Figure 21 One cycle of the polymerase chain reaction (PCR)

Knowledge check 27

How many lengths of double-stranded DNA are there after eight cycles of PCR?

Knowledge check 28

How many types of dNTP (deoxyribonucleoside triphosphate) are used in PCR?

There is a kit available for using PCR to amplify chloroplast DNA. Read the student's notes at:

www.ncbe.reading.ac.uk/ncbe/materials/DNA/plantevomodule.html

Follow the PCR process using animations here:

www.maxanim.com/genetics/PCR/PCR.htm

www.dnalc.org/resources/animations/pcr.html

Gene sequencing

You can follow an animation of this process at:

www.pbs.org/wgbh/nova/genome/media/sequence.swf

The sequence of steps involved in reading the sequence of nucleotides in a section of DNA is shown in Figure 22.

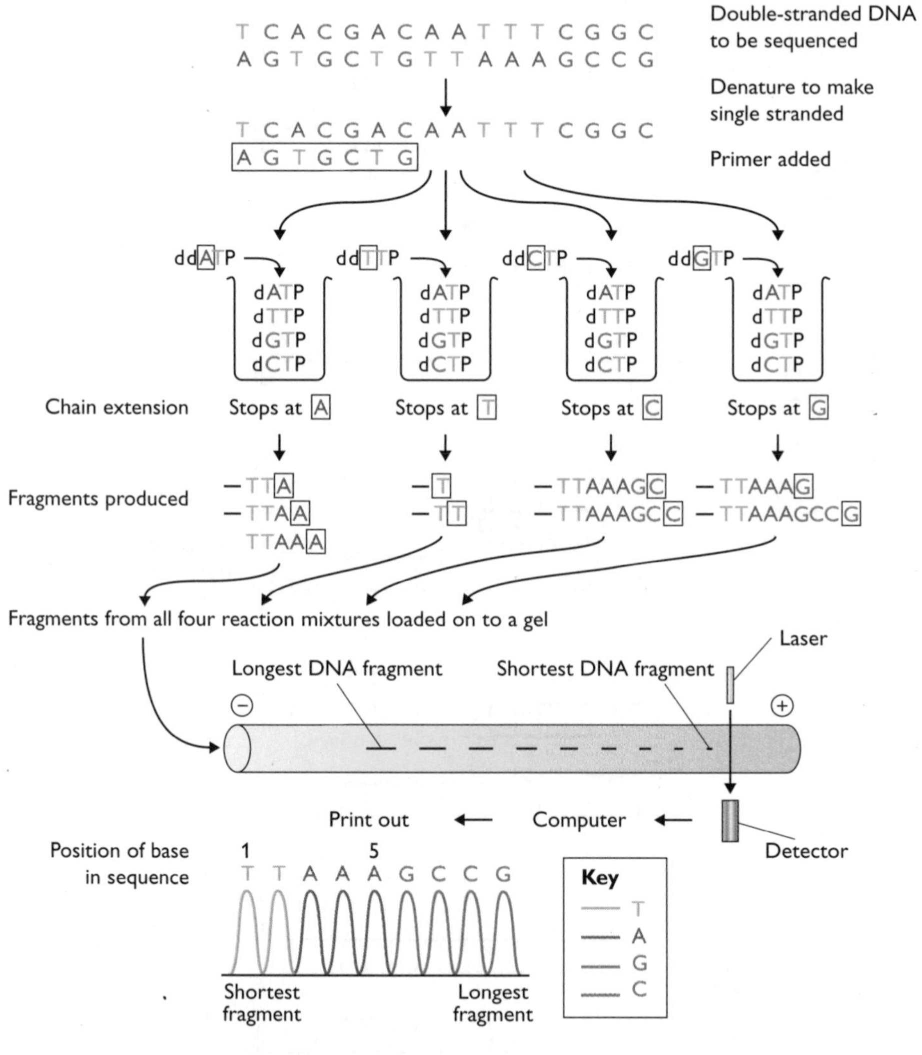

Figure 22 Gene sequencing using the chain-termination method

This method utilises DNA nucleoside triphosphates (dNTPs). These are used as they would be in DNA replication by DNA polymerase. The chains terminate because some nucleoside triphosphates are used that do not have a hydroxyl group, which means that they attach to a growing polynucleotide chain but cannot form a bond with another dNTP. Four types of dideoxynucleoside triphosphates (ddNTPs) are used with the bases A, C, T and G. Each of the four types of ddNTP is labelled with a fluorescent dye with a specific colour. These are mixed with normal dNTPs and added to reaction vessels as shown in Figure 22. When, by chance, these ddNTPs become part of the new chain, synthesis by DNA polymerase stops.

Fragments of newly synthesised DNA are produced which have different lengths. They are all mixed together and then separated by electrophoresis (Figure 20) in a capillary tube. As each fragment passes a laser and sensor the colour detected indicates the base at which replication terminated.

This technique gives the sequence of nucleotides (base pairs) in genes. Comparisons can then be made between individuals within a species to assess the variation at base-sequence level. It is also possible to compare species. We have found that many genes, such as the homeobox sequences of transcription factors (see p. 13) and the genes that code for β-globin in different animals (see p. 16), are similar in very different organisms. Homeobox genes have been conserved through evolutionary time because they code for proteins that bind to DNA and must always have the same tertiary structure and shape. Differences in gene sequences across the genome between species show how long ago they diverged (see Question 3, p. 83).

Genetic engineering

All DNA has the same type of structure. This means that pieces of DNA from one organism can be incorporated into the DNA of another (Figure 23). Cells that contain 'foreign' DNA — DNA from another organism — translate the code and make the same polypeptide. This sounds quite simple. However, often, when animal and plant genes are inserted into bacteria, although the correct sequence of amino acids is produced, the bacteria cannot fold and cut the polypeptide to produce a functional protein. Genes are transferred with promoter sequences so they will be expressed in the host organism.

Examiner tip
This section requires a good knowledge of the structure of DNA. Make sure you revise this thoroughly from Unit F212 as you may need to use terms like nucleotide, polynucleotide and the names of the bases when answering questions.

Stage 1 The gene for cloning is obtained by one of three methods.

(1) Restriction endonucleases cut DNA to give a staggered cut that leaves 'sticky ends' (some cut straight across the DNA to give blunt ends, in which case short sequences of nucleotides are added to give 'sticky ends').

Different restriction enzymes show specificity by cutting at different restriction sites. The enzyme shown (Eco R1) cuts at a site that is palindromic to give 'sticky ends':

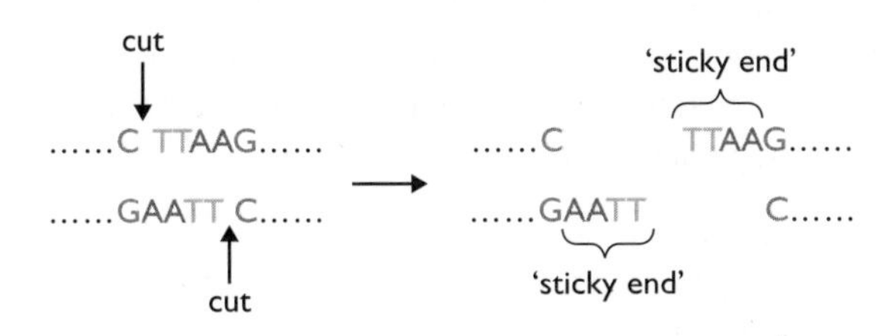

A gene (DNA) probe can be used to check that it is the correct DNA sequence. A gene probe is a small length of single-stranded DNA that can base pair with a base sequence on a longer section of DNA and so locate a specific (complementary) base sequence.

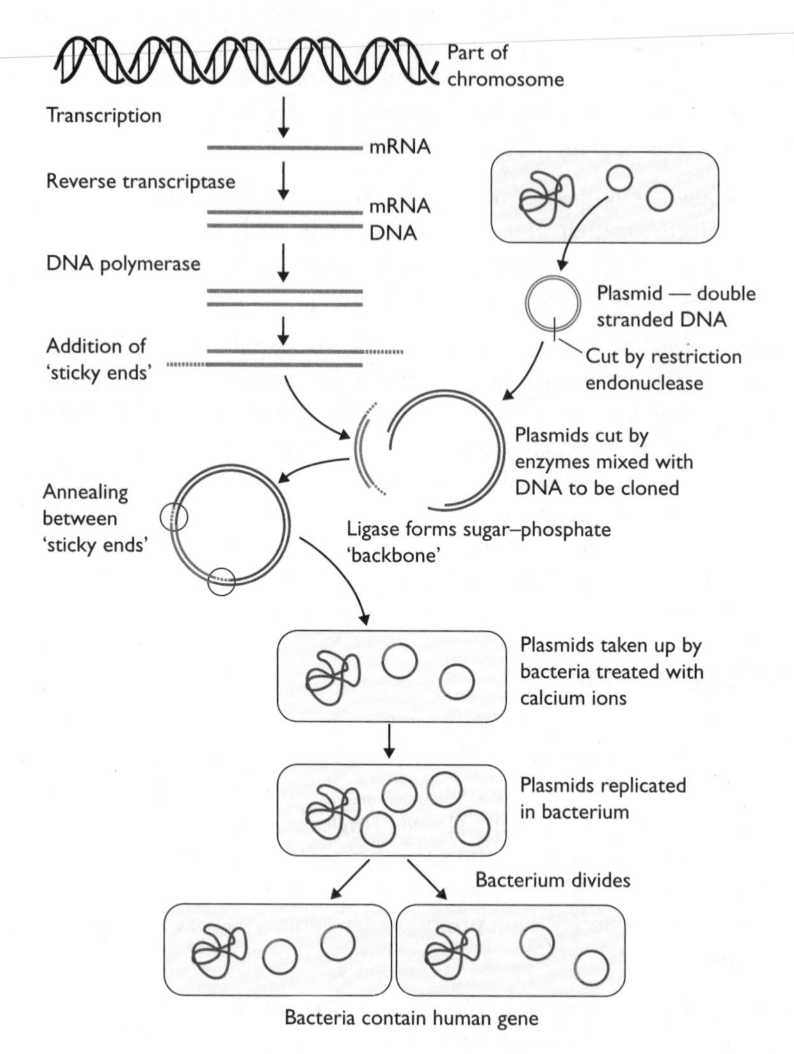

Figure 23 Gene cloning: using bacteria to make multiple copies of a human gene

(2) RNA is extracted from a cell. The enzyme reverse transcriptase makes a single-stranded copy of DNA using the RNA as a template. DNA polymerase makes a complementary polynucleotide (cDNA). This is how the gene for human insulin is obtained (see p. 34 of the Unit F214 guide).

(3) If the protein sequence is known, then a gene can be manufactured using a 'gene machine' that assembles nucleotides in the desired sequence using the information from the genetic code.

Stage 2 The gene is inserted into a vector.

(4) Multiple copies of the gene may be made by PCR.

(5) A plasmid is cut open using the same restriction enzyme.

(6) 'Open' plasmids and the required gene are mixed. Hydrogen bonding between the 'sticky ends' attaches the two — a process known as annealing.

(7) The enzyme ligase is added to form covalent phosphodiester bonds to join up the sugar-phosphate backbone of DNA. This is known as ligation.

Knowledge check 29

Distinguish between these two processes: annealing and ligation.

Recombinant DNA (rDNA) has now been produced (rDNA is DNA formed by combining DNA from two different sources).

Stage 3 Plasmids are taken up by bacteria.

(8) Bacteria are treated with calcium ions to increase the chances of plasmids passing through their cell walls.

(9) Plasmids enter bacteria, which are now described as **transformed** because they contain foreign DNA.

(10) DNA polymerase in bacteria copies the plasmids; the bacteria divide by binary fission so that each daughter cell has several copies of the plasmid.

(11) The bacteria transcribe and translate the foreign gene. The bacteria are described as **transgenic**.

Transformed bacteria have to be identified because some plasmids do not take up the genes and some bacteria do not take up the plasmids containing the foreign gene. This can be carried out by inserting genes for antibiotic resistance into the plasmid. However, a safer method is to use a gene for an enzyme that produces a protein that fluoresces in ultraviolet light.

Golden Rice™

White rice is a staple food for much of the world, but it lacks β-carotene which the body can use to make vitamin A. An estimated 100 million to 200 million children worldwide have vitamin A deficiency, which can lead to blindness, increased susceptibility to diarrhoea, respiratory infections, and childhood diseases such as measles. Rice synthesises β-carotene in its chloroplasts, but not in the endosperm as some key enzymes in the pathway are not expressed.

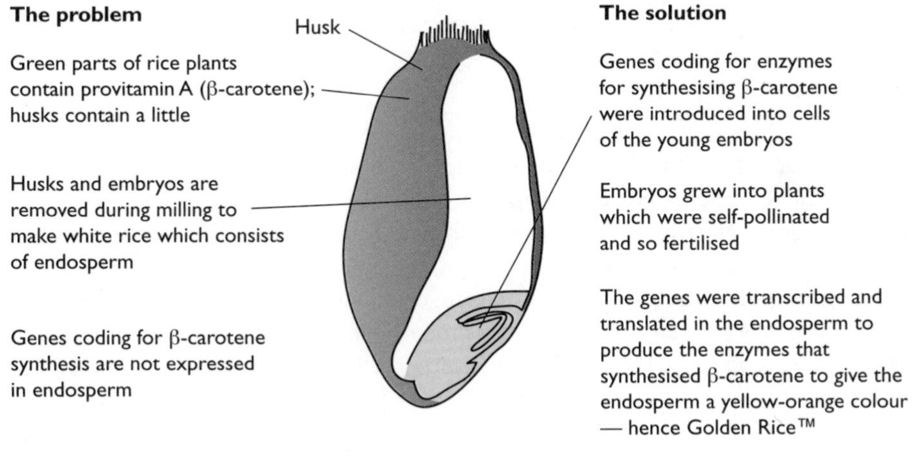

Figure 24 Golden Rice™

One way to genetically modify plants uses the bacterium *Agrobacterium tumefaciens*, which infects plants causing a cancerous disease called crown gall. It does not infect the monocotyledon group of plants that includes all the cereals (rice, wheat, barley, maize). Researchers in the Golden Rice™ project got around this by using *A. tumefaciens* to infect rice embryos. The bacterium has a tumour-inducing plasmid (T_i) that moves from the bacterium to plant cells to stimulate the host to make compounds that promote the growth of bacteria. Foreign genes can be inserted into the T_i plasmid that acts as the vector. Genes move from the plasmid into the chromosomes of the host's cells.

Other crops have been fortified in a similar way using traditional plant breeding techniques. Cassava is another crop that has been genetically modified to improve the supply of vitamin A to poor people in Africa.

Field trials have been carried out in the USA and the Philippines. It is possible that Golden Rice™ will become available to farmers in the next few years. At the time of writing (2012) it is not. You can find up to date information about this project at:

www.goldenrice.org

Xenotransplantation

One of the major problems with organ transplants is rejection. The immune system destroys transplanted organs unless there is a very close match. This means that transplant patients have to take immunosuppressant drugs, which can have serious side effects. Pigs have been genetically engineered in a number of ways to find out if they could be used as a source of organs for transplant to humans (xenotransplantation means transplantation from one species to another).

Pigs have been genetically modified by inserting human genes for:

- a cell-surface protein that stops the complement system (part of the immune system) from attacking transplanted organs
- cell-surface antigens (HLA antigens) used in cell recognition in humans
- an enzyme that regulates anti-apoptosis genes to prevent programmed cell death in transplanted organs

Pigs have also been genetically engineered to stop the expression of genes that produce cell-surface antigens which provoke the production of antibodies when pig tissue is transplanted into primates.

For xenotransplantation to be feasible it is necessary to breed transgenic pigs, which at present is not permitted in the UK.

Gene therapy

The transfer of a gene into an organism to repair a genetic fault and thereby treat or cure a genetic disorder is **gene therapy**. There are two forms:

- **Somatic gene therapy** — the gene is inserted into certain cells and cannot be passed onto the next generation.
- **Germ-line gene therapy** — the gene is inserted into a gamete or into a zygote so is therefore in every cell of the body. This means that it will be in the cells that give rise to the reproductive cells of that individual — the germ-line cells — and can be inherited.

In the case of recessive genetic disorders a dominant allele must be inserted into cells where it will be expressed. A good example is the treatment of SCID (severe combined immunodeficiency disease), which is a rare autosomal recessive disease. Children with this condition have no defence against common infections. Some were kept in sterile environments (sometimes known as a 'bubble') in the hope that treatments or a cure would be found. The non-functioning enzyme is adenosine deaminase (ADA) which is involved in the breakdown of purines, which are toxic to T cells. In 1990, T cells were removed from a girl who had SCID. They were given the normal allele for ADA using a retrovirus as the vector and the transgenic T cells were placed in her bone marrow. This treatment was successful and has been repeated on several children.

When the genetic disorder is caused by a dominant allele, as it is in Huntington's disorder, simply inserting the functional allele will not work because it is masked by the faulty allele. Another problem is that there are many different mutations that give rise to these conditions. A different approach has to be used. One possibility is to use RNA interference (RNAi) in which RNA combines with the mRNA from *both* the mutant allele and the normal allele. This makes double-stranded RNA, which cannot be translated. This technique effectively silences the genes because it stops the mRNA being translated. DNA for the normal allele is introduced but this is changed slightly so it produces a mRNA that does not combine with the RNAi and so is translated.

Although gene therapy holds great hope for curing genetic diseases, so far the success rate is low.

Germ-line therapy is not permitted in the UK and USA. It is illegal in the UK under the terms of the Human Fertilisation and Embryology Act 1990. However (as of 2012), it is considered likely that it may be used in the near future to correct disorders of mitochondrial DNA (mtDNA) that exists outside the nucleus. There are 150 such disorders that are passed on from mother to offspring in mtDNA.

It is unlikely that gene therapy could be used for polygenic conditions, such as coronary heart disease, stroke and diabetes in the near, or even distant, future.

Examiner tip

There are several different causes of SCID, ADA deficiency being one. Gene therapy has been used successfully to treat an X-linked form of the condition. SCID would make a good context for an exam question covering aspects of this unit, using synoptic material on immunity from F212.

Examiner tip

Inheritance from mother to offspring is known as maternal inheritance. This would make a good synoptic question for this unit as it could test your knowledge of mitochondria from F211 and F214.

Benefits and potential risks of genetic engineering

Microorganisms

Benefit — GM bacteria and yeasts produce human proteins that are otherwise in short supply, such as factor VIII (blood clotting factor), growth hormone and tissue plasminogen activator that helps to break down blood clots.

Potential risks include:

- GM bacteria and yeasts may escape into the wild and pass their foreign genes to pathogenic microorganisms, making the diseases the latter cause harder to treat.
- Antibiotic resistance genes in plasmids are often used as markers to identify transformed bacteria. This could reduce the effectiveness of these antibiotics in treating human and animal diseases.

Plants

Benefit — GM crop plants with herbicide resistance mean that herbicides can be sprayed during the growth of the crop to kill weeds without affecting the crop.

Potential risks include:

- Herbicide resistance genes may be transferred in pollen to species related to the crop. Weeds may become herbicide-resistant 'superweeds'.
- The increased use of herbicides could result in loss of biodiversity.

Animals

Benefit — transgenic sheep and goats make human proteins in their milk. These proteins are used to treat diseases such as hereditary emphysema.

Potential risk — insertion of foreign genes may have unforeseen effects.

Humans

Benefit — gene therapy has been used to treat SCID.

Potential risk — inserted genes may have unforeseen effects, such as disturbing the expression of other genes. If there are such problems, then germ-line therapy, if permitted, may have serious long-term consequences for future generations.

There are also ethical concerns over tampering with DNA of different species in ways that could never happen in nature. There are also some people who are sceptical of the intentions and procedures of the large biotechnological companies which, they say, are more interested in profits than in the long-term welfare of humans and the environment.

Examiner tip

To prepare yourself for questions on ethical issues surrounding cloning and genetic engineering read the genuine concerns of various groups and the counterarguments from scientists and others. Good places to start are: **www.beep.ac.uk** and **http://learn.genetics.utah.edu**

Synoptic links

This topic relies on your knowledge of the structure of DNA and RNA (pp. 21–24) in the Unit F212 guide), insulin (pp. 31–35) in the Unit F214 guide) and immunology (pp. 44–51) in the Unit F212 guide).

Summary

- Electrophoresis is the separation of negatively charged DNA fragments in a gel using an electric field. They separate by size.
- Sequencing a genome involves cutting chromosomes into sections and using ddNTPs in chain termination to give sections of bases that are then separated by electrophoresis.
- Gene sequencing allows comparisons between the genes of individuals of the same and different species.
- Recombinant DNA is composed of DNA from two or more different organisms, typically from different species.
- Genetic engineering involves extracting genes from one organism, or the manufacture of genes, in order to place them into another organism (often of a different species) so that the receiving (transgenic) organism produces the polypeptide(s) coded by the gene or genes.
- Restriction enzymes remove sections of DNA containing a desired gene by cutting across the sugar phosphate backbone at restriction sites composed of specific sequences of bases.
- The enzyme reverse transcriptase uses mRNA isolated from cells as a template to produce cDNA.
- DNA probes have base sequences complementary to desired DNA fragments; they are often made radioactive to locate them.
- The polymerase chain reaction (PCR) uses a heat-stable enzyme, *Taq* polymerase, to make multiple copies of DNA fragments.
- Plasmids, viruses and liposomes are used as vectors to insert DNA into organisms. DNA fragments are placed into plasmids using the enzyme ligase to form phosphodiester bonds between deoxyribose and phosphate.
- Plasmids are taken up by bacterial cells in order to produce a transgenic microorganism that can express a desired gene product. Calcium ions are used to increase the uptake of plasmids by bacteria.
- Genetic markers in plasmids, such as antibiotic resistance and fluorescent markers, are used to

identify bacteria that have taken up recombinant plasmids.

- Golden Rice™ is a genetically engineered rice that has genes that code for enzymes in the endosperm to make β-carotene, which is metabolised into vitamin A.
- Xenotransplantation is the use of animal organs to replace human organs. Pigs have been genetically engineered to express genes to reduce the chances of rejection if their organs are implanted into humans.
- Somatic gene therapy is the transfer of functioning alleles into cells to treat genetic conditions; germ-line gene therapy is inserting alleles into gametes or zygotes for the same reason.
- There are many ethical concerns raised by the genetic manipulation of organisms.

Ecosystems

Key concepts you must understand

An **ecosystem** is a place that consists of a community of organisms, the physical (abiotic) factors that influence it and all the interactions between organisms and between organisms and abiotic factors.

Ecosystems are dynamic, with energy flowing from the sun through autotrophs to heterotrophs and decomposers. Elements, such as carbon, nitrogen, sulfur and phosphorus, are continually cycled within ecosystems between the living organisms and the abiotic environment. Energy flow and nutrient cycling are influenced by human activities in artificial ecosystems, such as arable and livestock farms.

Examiner tip

You should revise the ecology that you studied in Unit F212 while learning the topics in this section. Expect to be asked about fieldwork techniques in the exam paper.

Key facts you must know

Food chains and food webs

A **trophic level** is a feeding level in a food chain. Producers are organisms that trap sunlight and make use of the energy in photosynthesis. The energy is used to convert the simple inorganic compounds, carbon dioxide and water, into complex organic molecules, such as sugars and amino acids. This provides biological molecules for the growth of autotrophs, such as plants, but also food for animals that feed on those plants. Some of the energy captured by plants is consumed and converted into animal tissue by herbivores. These may be ingested by other consumers at higher feeding levels (i.e. carnivores and humans). This pattern of energy flow is the **grazing food chain**, as in the following marine example:

Producer → Primary consumer → Secondary consumer → Tertiary consumer

Phytoplankton → Zooplankton → Small herbivorous fish → Large carnivorous fish

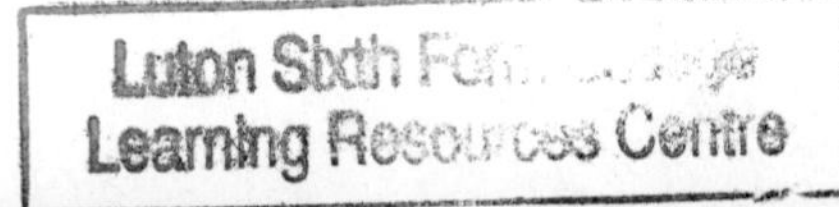

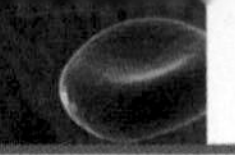

Table 12 Definitions of key terms in ecology

Term	Definition	Examples
Habitat	A place where an individual, population or community lives	Rocky shore; rock pool; pond; stream; meadow; forest; sand dune
Niche	The role of an organism in a community — where it is, what it does, how it feeds, how it behaves	A crab that lives in a rock pool scavenging on dead animal matter
Population	A group of individuals of the same species in the same area at the same time; males and females within the population can breed with each other	All shore crabs, *Carcinus maenas*, on a rocky shore
Community	All populations of plants, animals, microorganisms in a well defined area at the same time	All the organisms on a rocky shore
Ecosystem	A community and abiotic factors that influence it; interactions between organisms within the community	A rocky shore community and physical features that influence it
Abiotic factors	Physical and chemical factors that influence populations in a community	Wave action, temperature, pH, salinity, exposure to the air, CO_2 and O_2 concentration; ions; water supply
Biotic factors	Biological factors that influence populations in a community	Competition, predation, parasites (disease); grazing

Knowledge check 30

Draw grazing and decomposer food chains for a grassland ecosystem.

Detritus is dead organic matter that is a food source for **detritivores**, such as earthworms. **Decomposers**, such as bacteria and fungi, use dead organic matter from plants and animals as an energy source. This pattern of energy flow is the **decomposer food chain**. Energy is transferred from the grazing to the decomposer food chain in the form of dead leaves, faeces and, after death, animal tissue.

Food chains tend to be short. Much of the energy that enters the organisms at one trophic level is used by those organisms and is, therefore, not available to the next trophic level. Energy is lost during respiration and in heat loss. This means that only a small percentage of the energy that enters a trophic level becomes stored in the bodies of the organisms in that trophic level to be eaten by those of the next.

Food webs show all the feeding relationships within an ecosystem. The arrows in a food web show the direction of energy flow.

Energy is not recycled — it leaves the ecosystem as infrared radiation that warms the atmosphere. A moment's thought should tell you that it cannot be recycled as sunlight! The flow of energy between trophic levels is not very efficient, as is outlined below.

Energy flow from plants to primary consumers

Little of the light energy that strikes plants is used in photosynthesis. Reasons for this include:

- light is reflected from the surfaces of leaves
- light passes straight through leaves
- it is too cold for the chloroplast enzymes to function efficiently
- carbon dioxide is in short supply

The **gross primary productivity** is the energy trapped by plants in photosynthesis per unit time. Some of this is used in respiration. The remainder, that is available for maintenance and new growth in the plants, is the **net primary productivity**. At best, crop plants make available to us 5% of the energy that strikes their leaves. In natural ecosystems, the percentage of energy available to primary consumers is even lower. Not all of the energy in plants reaches the primary consumers as some plant matter is not eaten, some cannot be digested and much will die and decay rather than be eaten by grazers.

Energy flow from primary consumers to secondary consumers

Some animals, for example antelope, feed on plants. Their energy input is equivalent to the energy content of all the grass and other plants that they eat. Some animals are preyed on by predators, such as lions, cheetahs and leopards. The energy in the antelope that is eaten by a predator is the energy transferred to the next trophic level. At best, about 10% of the energy entering the primary consumer trophic level is passed to the secondary consumer trophic level. In this example, reasons for this include:

- Antelope use energy in keeping warm (especially at night).
- They move about in search of food.
- Energy is lost as heat during digestion of their food.
- Not all the antelope's body is eaten.
- Not all of the antelope eaten by the predator is digested and absorbed.
- They use energy in producing gametes and mating; they also use energy in developing and rearing young.
- Respiration is about 30% efficient at transferring energy to ATP.

The only energy transferred to the next trophic level is energy in new flesh. As a percentage of the energy input from the producers that is very small.

In studies of natural and artificial ecosystems ecologists take samples, find the energy content of biological materials, examine faecal contents and gut contents. This gives an idea of the quality and quantity of material produced by autotrophs and eaten by heterotrophs. The efficiency of primary producers is calculated from the light energy striking the plants and from the energy they trap in the products of photosynthesis.

Ecological efficiency is the efficiency of energy transfer between trophic levels. It is calculated by comparing the energy consumed by a trophic level with the energy consumed by the next trophic level. Between secondary and tertiary consumers, this is calculated as follows:

$$\frac{\text{energy consumed by tertiary consumers}}{\text{energy consumed by secondary consumers}} \times 100\%$$

Arable farmers and growers attempt to make net primary productivity as high as possible. Livestock farmers do the same for secondary productivity. Table 13 shows how human activities can alter the efficiency of energy flow.

Table 13 Some ways in which humans manipulate energy flow in artificial ecosystems

Method	Crop plants (producers)	Livestock (primary consumers)
Maximise energy input	Optimum planting distances between crop plants Provide light for greenhouse crops on overcast days	Provide good quality feed

Examiner tip
Remember 'energy is neither created nor destroyed'. Energy cannot be 'lost' and disappear completely. If you write 'energy is lost...' you mean it is transferred into a form that is of no use to organisms.

Examiner tip
The energy content of biological materials is determined by burning them in oxygen within a piece of apparatus known as a calorimeter. The increase in temperature is recorded and from this the energy content in joules is calculated. This section is an opportunity for questions that test your maths skills.

Method	Crop plants (producers)	Livestock (primary consumers)
Maximise growth	Provide water (irrigation); fertilisers (containing NPK and other elements, e.g. S) Selective breeding for fast growth	Provide food supplements, e.g. vitamins and minerals Selective breeding for fast growth
Control disease	Fungicides	Antibiotics and vaccines
Control predation	Fencing to exclude grazers, e.g. rabbits, deer Use pesticides to kill insect pests, nematodes, slugs, snails, etc.	Extensive systems (ranching): control predators such as wolves and foxes Intensive systems: keep animals protected from predation in sheds
Reduce competition	Ploughing and herbicides kill weeds	Control competitors such as rabbits and deer
Reduce energy loss	Breed plants that maximise energy storage in edible products, e.g. seeds and fruits	Keep animals in sheds — less energy lost in movement and maintaining body temperature

Succession

When ground becomes available for colonisation for the first time a process of **primary succession** begins in which the community changes over time. The sequence of communities that develops is known as a **sere**. The type depends on the nature of the environment; for example the type that occurs on a sand dune is a psammosere.

Examiner tip

In the Unit guide for F213/F216, you will find some examples of fieldwork exercises that will give you some help with this part of the specification. See examples 34–38, which give information about assessing the distribution and abundance of species using frame quadrats, line transects and belt transects.

Plants and animals in different seral communities show adaptations to survival. Those in pioneer communities show adaptations to low nutrient and water availability and harsh abiotic factors. Pioneer plants often self-pollinate because individual plants are widely scattered.

Most climax communities are dominated by large trees. The mix of species is determined by abiotic factors, such as climate, altitude and soil type. Productivity is low since trees have much woody biomass that does not carry out photosynthesis. Earlier stages in the succession are much more productive.

Decomposition and nutrient cycling

Detritivores are organisms that ingest dead matter (detritus) and shred it, increasing its surface area. Examples of detritivores are earthworms, termites and woodlice. The material they egest in their faeces has a large surface area, so providing easy access for decomposers such as bacteria and fungi. Decomposers secrete enzymes that catalyse the hydrolysis of large, insoluble organic molecules into small, soluble molecules that they absorb through carrier proteins using facilitated diffusion and active transport; the substances are then used in metabolism. Carbon in these compounds is lost to the environment as carbon dioxide when the compounds are respired. This 'unlocks' carbon from dead organic matter, making it available for carbon fixation in photosynthesis.

Knowledge check 31

Explain the importance of detritivores and decomposers in ecosystems.

Detritivores and decomposers show **saprotrophic** nutrition as they feed on dead organisms.

Nitrogen is an important element because it is part of many biological molecules, including:

- amino acids and proteins
- nucleotides and nucleic acids
- most phospholipids

Most organisms take in nitrogen that is already 'fixed' in that it is combined with another element such as hydrogen or oxygen. Plants absorb simple forms of fixed nitrogen, such as ammonium ions or nitrate ions. Animals obtain nitrogen by eating food containing complex nitrogenous compounds, such as proteins. There is only a limited supply of 'fixed' nitrogen in natural ecosystems and the supply of nitrate ions for plants depends on the action of microorganisms that use nitrogen compounds. Part of the nitrogen cycle is shown in Figure 25.

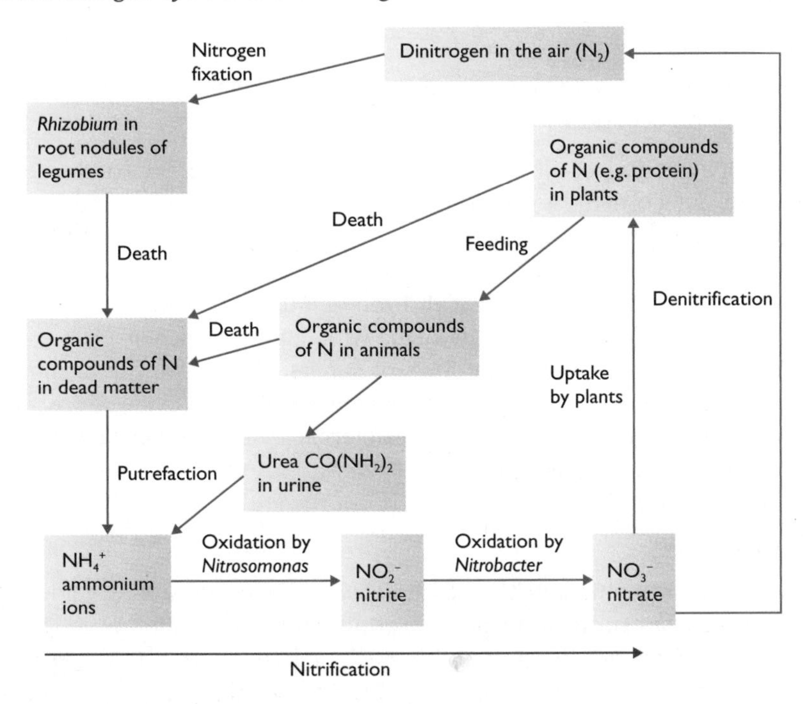

Figure 25 Part of the nitrogen cycle (note the importance of bacteria in recycling nitrogen compounds)

Use Figure 25 to find the features of the nitrogen cycle that follow.

Features involving plants:

(1) Absorb nitrate ions (NO_3^-) from the soil
(2) Convert nitrate to ammonium ions (NH_4^+) and use them to make amino acids by the process of amination (remember that amino acids have the amino group $-NH_2$)
(3) Assemble amino acids into polypeptides and proteins
(4) Lose leaves and die, forming dead matter (e.g. leaf litter) that contains proteins

Features involving herbivores:

(1) Eat plants and digest proteins to amino acids
(2) Use amino acids to make proteins (e.g. haemoglobin, collagen)
(3) Break down excess proteins and amino acids to form ammonium ions

(4) Convert ammonium ions to urea in the ornithine cycle
(5) Excrete ammonium ions and urea in urine
(6) Produce dung that contains proteins
(7) Die, leaving bodies that contain proteins

Features involving putrefying bacteria:
(1) Digest proteins in dead organisms into amino acids
(2) Deaminate amino acids to form ammonium ions

Features involving nitrifying bacteria:
(1) *Nitrosomonas* oxidises ammonium ions to nitrite ions to provide itself with energy
(2) *Nitrobacter* oxidises nitrite ions to nitrate ions to provide itself with energy
(3) Nitrate ions released by *Nitrobacter* are available for plants to absorb and so the cycle is completed

Knowledge check 32

Make a table to show the roles of microorganisms in the nitrogen cycle.

Examiner tip

To help learn the nitrogen cycle, make a large flow chart and use it to include all the synoptic links that you can identify. Nitrogen is the key: think of all the biological molecules that contain nitrogen and identify their roles in organisms.

Features involving denitrification:
(1) Nitrate ions are converted into dinitrogen (N_2) by denitrifying bacteria
(2) Denitrification occurs in waterlogged, anaerobic soils decreasing their fertility.

There is a huge quantity of nitrogen in the atmosphere. About 80% of the air around us is nitrogen, but it is in the form of the gas, dinitrogen (N_2), in which there is a strong triple bond between the nitrogen atoms, ($N \equiv N$). Some bacteria, such as *Rhizobium*, fix nitrogen by using energy (from ATP) and the enzyme nitrogenase to break the triple bond and combine nitrogen atoms with hydrogen to form ammonium ions (NH_4^+). *Rhizobium* lives in a mutualistic relationship with legumes, such as peas and beans, which provide it with energy in the form of sugars. The active site of nitrogenase also accepts oxygen, so legumes have leghaemoglobin to provide *Rhizobium* with anaerobic conditions in the nodules so the bacteria can fix nitrogen. Leghaemoglobin absorbs oxygen, keeping it away from the bacteria.

Synoptic links

The basis of autotrophic nutrition is photosynthesis which was covered in Unit F214. Energy losses occur in respiration (Unit F214). Respiration is only 30–40% efficient in transferring energy from organic compounds to ATP. The rest is lost heating the body and/or the surroundings. It is only energy trapped in biological molecules eaten by the next trophic level that is passed up the food chain. Different respiratory substrates have different energy values (p. 70 of the F214 guide).

Nothing works better than the nitrogen cycle to bring together different aspects of biology. The nitrogen cycle has trophic levels, food chains, different ways of feeding, biological molecules (amino acids and proteins), nutrition of organisms, enzyme action (nitrogenase) and mutualism (*Rhizobium* and legumes), and much more.

Populations and sustainability

Figure 19 shows a sigmoid growth curve for a population of bacteria in a simple ecosystem with no competitors, predators or parasites. Similar growth occurs when an organism is released into a new environment where there are few or no factors to limit its growth. Animals such as rabbits, rats, goats, cats and mice show this when introduced into new environments. The cane toad, introduced into Australia from

Hawaii (to where it had been introduced from South America), is a good example. The time span is longer than in Figure 19 but the pattern is the same. Growth slows eventually because limiting factors provide **environmental resistance** and the population has reached its **carrying capacity** — the maximum number of individuals that an environment can sustain. Many populations rise to a maximum and then fluctuate about a mean. Some populations show J-shaped growth curves with a population explosion followed by a crash when resources run out because the carrying capacity is exceeded.

Two limiting factors for population growth are competition for resources and predation.

Intraspecific competition is competition between members of the same species. This is intense because individuals require the same resources and have the same methods for obtaining them.

Interspecific competition is competition between individuals of different species. This can happen when an organism migrates, or is introduced, into a new environment. It is likely that there are no available niches and so there is competition between the invading species and the existing species. This has happened in New Zealand with the introduction of rats that have displaced ground-dwelling birds, such as the kiwi.

Predators and prey

The relationship between predators and their prey has been studied in simple ecosystems and in the laboratory in conditions where the predator had only one prey species and the prey species had only one predator (see Question 5, p. 92). The standard textbook example is the Arctic hare and snowshoe lynx. It is assumed commonly that the population of the prey species is controlled by the predator, but it is far more likely that the effect of predators on prey numbers is small. The quality and quantity of plant food available to the prey species controls its numbers, and the prey species' numbers in turn control those of the predator. Many small mammal species show 10-year cycles determined by the availability of food plants. It is these cycles that probably have the most effect in controlling the populations of their predators.

Conservation

Many natural and man-made ecosystems are at risk from human activities. People often think that natural and man-made ecosystems should be preserved for future generations. They confuse **preservation**, which involves keeping ecosystems and species *as they are now*, with **conservation**, which involves reclamation and management. Conservation recognises that ecosystems are dynamic and require understanding and managing, rather than simply preserving. It also involves proactive methods to maintain or increase biodiversity.

Timber production in the UK

Tree felling

Clear felling removes all the trees from a large area at the same time. This has adverse effects on the environment as the organisms living in that area either die or have to migrate, so reducing biodiversity. It also encourages soil erosion especially if the plantation is on a hillside. To prevent this, **strip-felling** is carried out by cutting down strips of trees thus avoiding the massive destruction of clear felling. **Selective felling** is employed to take out single trees.

Coppicing

After reaching a certain size, the trunks of some trees — for example, sweet chestnut, hazel and willow — are cut down to a stump. New shoots grow up from the stump and, after a few years, these can be harvested. Coppiced woodland is a biodiverse ecosystem. If different areas of the woodland are cut each year (rotational coppicing) the wood can be harvested without loss of biodiversity because the ground flora does not lose the protection of the trees. This also allows more light to the ground plants that would become extinct if the trees grew to full size with a dense canopy of leaves.

Knowledge check 33

Explain why timber production in the UK may be considered to be sustainable.

The Galápagos Islands: a case study in conservation

Charles Darwin arrived in the Galápagos in the Pacific on HMS *Beagle* on 15 September 1835. His time on the islands provided him with much evidence of endemic species — species found there and nowhere else. On the islands there are now:

- 560 plant species of which 180 are endemic
- 19 endemic reptile species, although a bright pink land iguana has been found recently which may make 20
- 29 endemic bird species including the 13 species known as Darwin's finches

The Galápagos Islands are a laboratory for studying speciation. Much information has been collected on Darwin's finches by scientists over the past 40 years (see Figure 15 on p. 29 for an example). Isolation and extreme selective pressures (such as drought) have brought changes to the bird populations.

The islands have been affected severely by human activities. In the past, whalers killed huge numbers of marine mammals; people have introduced goats, pigs, cats, rats, mice, dogs and many plant species, none of which was originally on the islands. These introductions compete effectively with indigenous species and are also grazers and predators. For example, elephant grass, *Pennisetum purpureum*, competes with the endemic daisy tree, *Scalesia pedunculata*.

Only four of the islands are inhabited but the population has increased greatly as tourism has increased. Parts of the archipelago are a national park (established in 1959 by the government of Ecuador) and the Charles Darwin Research Station carries out research and runs conservation projects. Some of the current threats to the wildlife include:

- the fast growth of the town of Puerto Baquerizo Moreno on the island of San Cristóbal
- an increase in rubbish and sewage — disposal creates environmental problems
- land is required for housing and agriculture
- pressure on drinking water and energy supplies
- high unemployment — many people have resorted to fishing for sea cucumbers and lobsters, so disturbing the marine environment

Steps taken to reduce these effects include:

- restrictions on visiting the uninhabited islands, such as Daphne Major
- limiting the areas on these islands that can be visited
- a marine reserve of 133 000 km^2 that protects against depredations by fishermen; this is cared for by local people as well as by conservation organisations
- culling alien species, such as goats, and destroying invasive plants, such as elephant grass
- captive breeding and reintroduction, e.g. of giant tortoises
- inspecting visiting boats for alien species to prevent colonisation

Synoptic links

Mockingbirds, *Nesomimus* spp., interested Darwin and prompted his thoughts about the development of new species. There is more about these birds at:

www.homepage.villanova.edu/robert.curry/Nesomimus/index.html

You can read more about conservation on the Galápagos at:

www.savegalapagos.org
www.darwinfoundation.org/english/pages/index.php

Summary

- An ecosystem is a dynamic system consisting of a community of organisms, the environmental factors and all the interactions between organisms and their environment.
- Biotic factors are interactions between organisms, such as competition within species (intraspecific) and between species (interspecific) and predation. Abiotic factors are physical factors that influence communities, such as temperature and pH.
- Light energy is transferred into chemical energy by producers (green plants and some prokaryotes) making energy available to consumers and decomposers. These are trophic levels in a food chain.
- Energy transfers between trophic levels are determined by comparing the energy taken in by one trophic level with the energy available to the next.
- Farmers and growers use techniques such as applying fertilisers and using protected environments to improve the flow of energy through artificial ecosystems.
- A primary succession is the sequence of changes from bare ground to a climax community, for example from a sand dune to woodland.
- Line transects, belt transects, quadrats and point quadrats are used to study the distribution and abundance of organisms.
- Decomposers — bacteria and fungi — break down organic material into simple inorganic compounds such as carbon dioxide and ammonia.
- Ammonia released by decomposers is oxidised by nitrifying bacteria to nitrite ions and then to nitrate ions. Denitrifying bacteria convert nitrate ions to dinitrogen (N_2). Nitrogen-fixing bacteria fix dinitrogen as ammonia.
- The carrying capacity is the maximum number of a species that can be supported by an ecosystem. Biotic and abiotic factors limit the final size of populations.
- The relationships between predators and their prey determine the sizes of their populations.
- Conservation, unlike preservation, is a dynamic process that involves managing ecosystems for the benefit of the long-term survival of species and sustainable production of natural resources, such as timber.
- There are economic, social and ethical reasons for conservation of biological resources.
- Human activities, such as fishing and the introduction of alien species, have had harmful effects on animal and plant populations in the Galápagos Islands.

Responding to the environment

Plant responses

Key concepts you should know

In plant communication systems there are several elements, including:

- a receptor to detect stimuli (changes in the environment)
- plant hormones
- an effector to bring about change

A seed germinates. The embryo inside grows to form a shoot system and a root system. A seed can be orientated in any direction in the soil, so it needs a mechanism to detect gravity so roots grow downwards and shoots grow upwards. Shoot systems need to respond to the direction of light so that leaves expose the maximum surface area to the light. Plants respond to stimuli such as gravity, direction of light, predation (grazing), water stress, changes in day length and temperature. This is to avoid the stress caused by adverse environmental (or abiotic) conditions.

Deciduous trees lose their leaves at the beginning of a very dry or very cold period when water is not readily available because there is little rainfall or water is frozen in the ground. As with responses to gravity and light, leaf abscission is controlled by hormones.

Auxins, gibberellins, cytokinins, abscisic acid (ABA) and ethene are plant hormones (also known as plant growth regulators or PGRs). There are natural and synthetic examples of each type.

Key facts you must know

Tropisms

Directional plant growth responses are called **tropisms**. Shoots grow towards light (positive phototropism) and away from gravity (negative geotropism).

Phototropins are cell surface membrane proteins that are phosphorylated when illuminated by blue light (see Figures 6 and 7 in Question 6, p. 95). In the phototropic response (Figure 26) they act as receptors that cause more of the auxin indole acetic acid (IAA) to accumulate on the shaded side of the shoot. Auxin stimulates cells to pump hydrogen ions into the cell wall. This acidification loosens the bonds between cellulose microfibrils and the surrounding matrix. Turgor pressure causes the wall to stretch more lengthways so these cells elongate, causing bending towards the light.

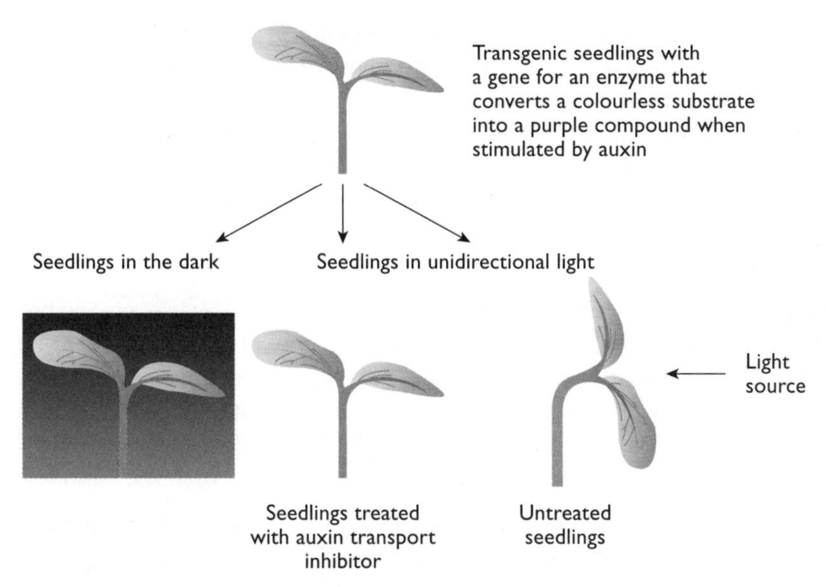

After 12 hours:
- seedlings immersed in colourless substrate
- increase in length of cells determined

The results are shown in the table.

	Seedlings in the dark	Seedlings in unidirectional light	
		Treated with auxin transport inhibitor	Untreated
Distribution of purple colour	Uniform through stem	Uniform through stem	More on shaded side
Auxin distribution	Uniform through stem	Uniform through stem	More on shaded stem
Increase in length of cells	All cells increased by same length	All cells increased by same length	Cells on shaded side increased more than an illuminated side

Figure 26 Transgenic seedlings were used to show that unidirectional light causes an unequal distribution of IAA in shoots

Apical dominance

The stem apex produces IAA which passes down the shoot. Removal of the tip causes lateral buds to grow into side shoots. Cytokinins are required to stimulate the growth of meristem cells in the lateral buds and it is thought that interaction between the two plant hormones ensures that lateral buds do not grow until the concentration of auxin decreases with distance from the shoot tip. The advantage of apical dominance is that it allows plants to direct most of their assimilates to the apical bud to give growth in height. This is important in gaining height over competitors and absorbing light.

Examiner tip

You learnt about assimilates in Unit F211. They are compounds such as sucrose and amino acids that are synthesised by plants.

Gibberellins and height

Plants that are genetically dwarf, such as some pea plants, are homozygous for genes that code for enzymes in the pathway that produces gibberellin. The gene, **Le/le**, codes for the last enzyme in the pathway. The dominant allele, **Le**, codes for

Examiner tip
DELLA stands for the first five amino acids in the primary sequence of these proteins. It is not a well-known abbreviation, but it is one you can use in an answer without having to explain it.

a functioning enzyme permitting normal gibberellin synthesis and resulting in the 'tall' phenotype. A substitution in the gene gives rise to a change from alanine to threonine in the primary structure of the enzyme near its active site. This mutation has given rise to the recessive allele, **le**. Homozygous plants, **lele**, are genetically dwarf. Gibberellins activate genes by inhibiting repressors. They do this by causing the breakdown of DELLA proteins that inhibit transcription factors

Abscission

Abscission is the term given to shedding of plant parts such as leaves. An abscission layer of cells forms in the petiole (leaf stalk) and the walls between them begin to break down, catalysed by the enzymes cellulase and pectinase which hydrolyse cellulose fibres and pectins in the cell wall matrix. Cells either side of the abscission layer develop protective waxy cell walls, and xylem and phloem are sealed by waxes. The leaf becomes loosely attached to the stem and falls off when the wind blows.

Plant hormones control the **senescence** (ageing) of leaves and their fall. Cytokinins decrease in concentration and this leads to fewer nutrients reaching the leaves. Auxin concentration decreases in the leaf and this leads to an increase in ethene which inhibits auxin production and stimulates the changes leading to abscission.

Table 14 Commercial uses of some plant hormones

Plant hormone	Example	Commercial use
Auxin	NAA	Rooting compound — stimulates root growth on stem cuttings; thinning of fruit
	2,4-D	Weed killer — stimulates rapid elongation growth to kill broad-leaved weeds in cereal crops
Gibberellin	GA_3	Promotes the growth of some fruit crops (e.g. grapes); increases yield of sugar in sugarcane; stimulates germination of barley in malting (start of brewing process)
Cytokinin	Benzyladenine	Improves fruit quality; prolongs storage life of green vegetables, such as asparagus, broccoli and celery
	Kinetin	Used in tissue culture as it stimulates cell division in buds to give shoot growth

Synoptic links

A synoptic question may ask you to compare hormones in flowering plants with animal hormones (from F214). To prepare for such a question you should make a table to compare the two groups of hormones using the following features: sites of synthesis; method of transport; modes of action; and effects (long and short term).

Animal responses

Key concepts you must understand

Animals must respond to their environment to obtain food and water, escape from predators, move to suitable conditions and shelter from adverse conditions. They have a sensory system to detect stimuli and two communication systems — the nervous system and the hormonal system — and the muscular and skeletal systems act as effectors to bring about responses.

Key facts you must know

The organisation of the mammalian nervous system is shown in Figure 27.

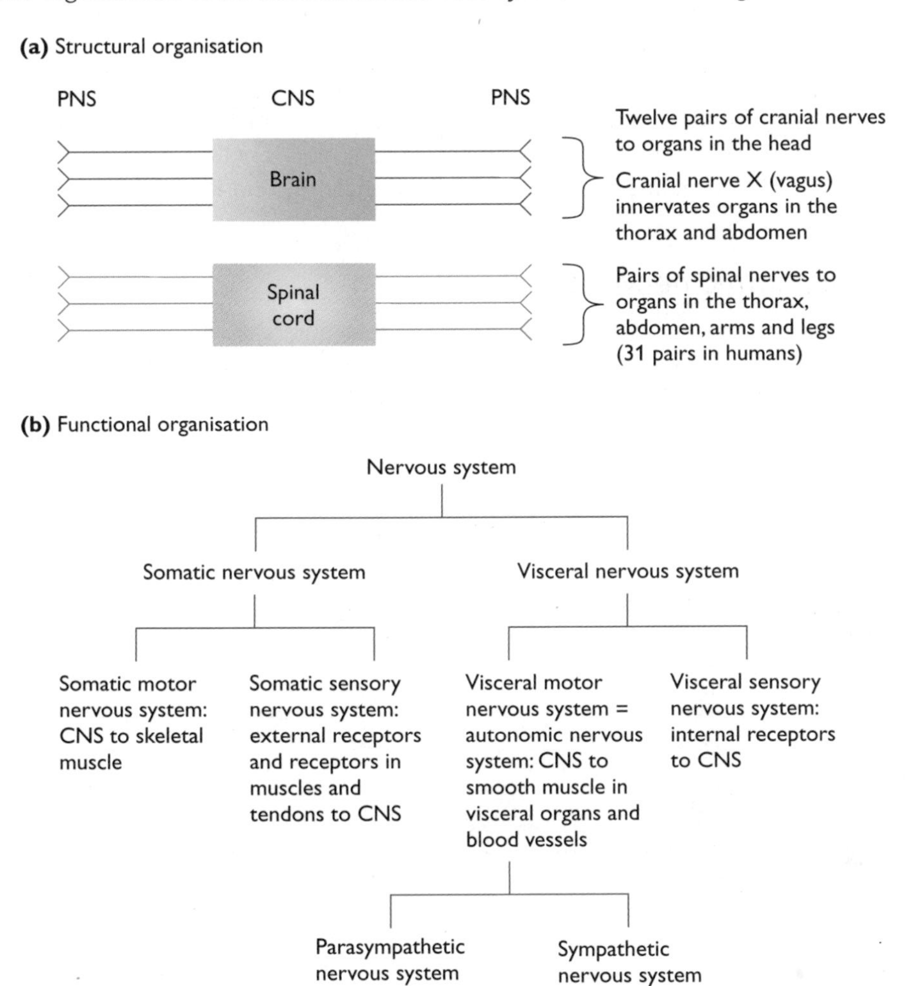

Figure 27 The organisation of the nervous system in mammals

Table 15 shows the three types of muscle tissue in vertebrates. The name used in this guide is given first, followed by other names in brackets.

Table 15 Types of muscle

Type of muscle	Cell structure	Distribution	Innervation
Smooth (involuntary, visceral, non-striped, non-striated)	Uninucleate cells; myofibrils not organised	Tubular organs of the viscera such as the gut (oesophagus to anus), airways, Fallopian tubes, uterus, arteries, arterioles and veins	Visceral sensory Autonomic: parasympathetic, sympathetic

Type of muscle	Cell structure	Distribution	Innervation
Striated (voluntary, skeletal, striped)	Multinucleate (many nuclei in a mass of sarcoplasm); myofibrils organised in parallel bundles	Muscles attached to the skeleton by tendons	Somatic sensory Somatic motor
Cardiac	Uninucleate cells joined by intercalated discs; myofibrils similar to striated	Heart	Visceral sensory Autonomic: parasympathetic, sympathetic

Examiner tip
You should expect questions on the nervous system that require your knowledge and understanding of conduction of impulses from Unit F214. Write about *impulses*, not 'signals' or 'messages'.

Smooth muscle and cardiac muscle are composed of individual cells. The biceps and triceps in the arm and the quadriceps femoris in the leg are examples of muscles composed mainly of striated muscle, which is not composed of individual cells. Each muscle 'cell' or fibre is a syncitium with many nuclei in one mass of sarcoplasm (muscle cytoplasm).

The autonomic nervous system

This system consists of all the motor neurones that innervate organs that are not usually under conscious control. The neurones innervate smooth muscle in the viscera and the heart, which receives dual innervation by:

- cardiac accelerator neurones that increase heart rate
- cardiac decelerator neurones that decrease heart rate

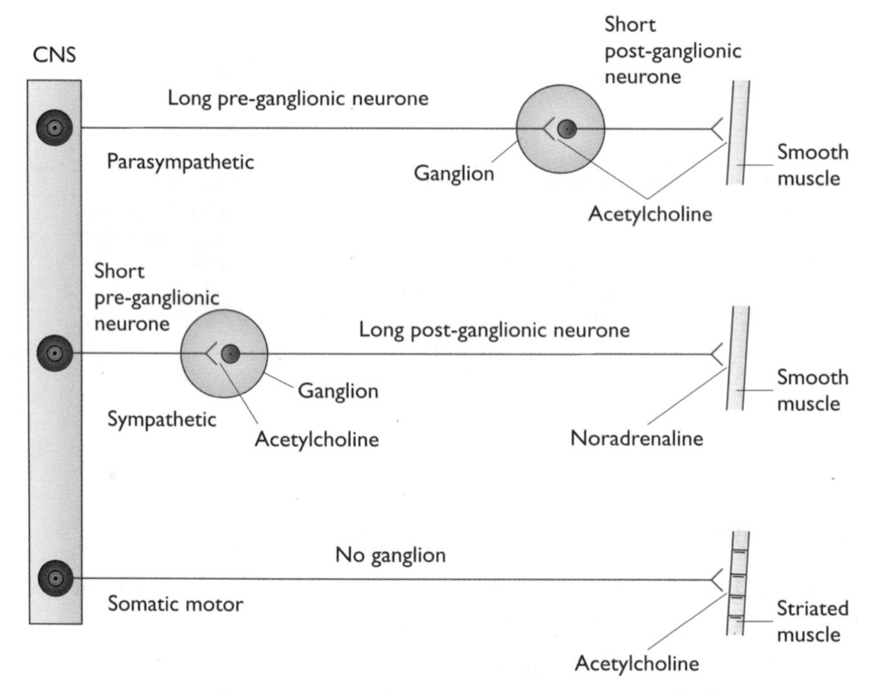

Figure 28 Neurones in the autonomic nervous system compared with those in the somatic motor system

Examiner tip
You learned about the control of the heart in Unit F211. This is a good opportunity to revise this topic as information on the sino-atrial node, atrioventricular node and Purkyne fibres may be expected in answer to a question in the exam paper for this unit.

The accelerator neurones are part of the **sympathetic nervous system** which tones up the body in response to stressful situations. This system works in conjunction with the catecholamine hormones adrenaline and noradrenaline. The decelerator neurones are part of the **parasympathetic nervous system** which tones down the body and helps to conserve resources.

In the **somatic motor system**, the cell bodies of neurones are in the brain or spinal cord and impulses pass directly to the muscles. Neurones in the autonomic nervous system are different: there are two neurones in series between the CNS and each effector. The synapses between the neurones are in swellings known as ganglia. You may read that the two systems act antagonistically. This is true of the control of the heart, but it is not true for all the aspects of physiology (see Table 16).

Table 16 Some of the functions of the autonomic nervous system

Organ/tissue	Sympathetic innervation	Parasympathetic innervation
Eye: iris radial muscle Eye: iris circular muscle	Contracts to dilate pupil Does not innervate	Does not innervate Contracts to constrict pupil
Adrenal medulla	Secretion of adrenaline	Does not innervate
Salivary glands	Small quantities of viscous saliva	Large quantities of dilute secretion rich in amylase
Alimentary canal	Inhibits, e.g. stimulates arterioles to contract (vasoconstriction), diverting blood elsewhere	Activates, e.g. stimulates muscle contraction and secretion from gastric glands and pancreas
Heart	Increases heart rate	Decreases heart rate
Whole body	Coordinates response to stress including mobilising resources	Coordinates conservation of resources

Structure and functions of the human brain

Figure 29 shows views of the human brain with the functions of the major parts.

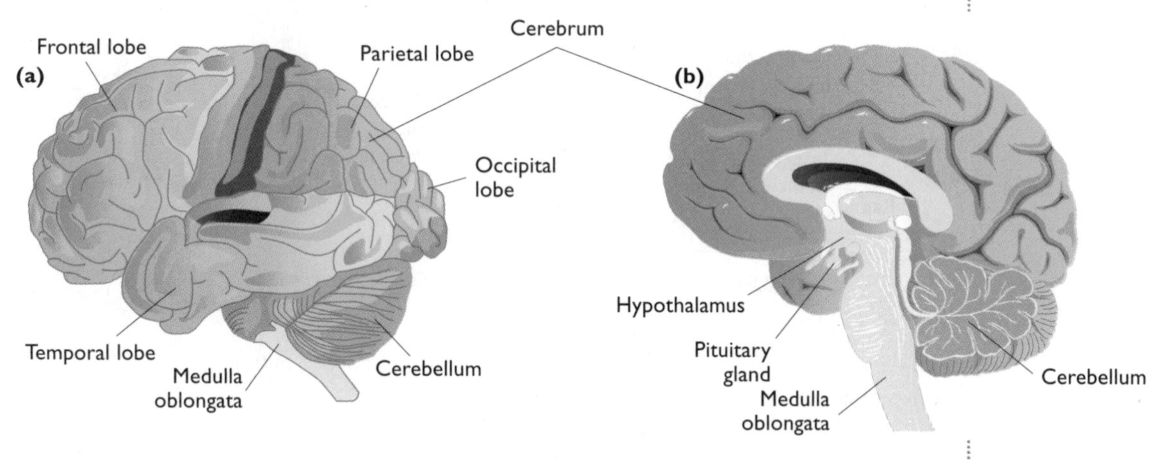

Region of brain	Functions	
Cerebrum	Conscious thought Coordination of voluntary activities Memory Association of incoming information with past experience	Learning and reasoning Understanding of language Control of speech Interpretation of visual, auditory and other external stimuli

Region of brain	Functions
Cerebellum	Interpretation of sensory input from muscles and tendons Coordination of balance, posture and movement Muscle coordination
Hypothalamus	Control of core body temperature, osmoregulation (water potential of blood), reproduction through anterior pituitary, secretion of hormones, e.g. ADH
Medulla oblongata	Regulation of autonomic activities, e.g. heart rate, blood pressure, breathing rate and peristalsis

Figure 29 (a) An external view of the human brain (b) A vertical section with the functions of the major parts

The control of muscular movement

The brain, spinal cord and peripheral nervous system work together to control the contraction of striated muscle. Neurones do not work in isolation — the simplest way in which they work together is the reflex arc. Movement of the forearm involves neurones that stimulate striated muscle in the biceps and brachialis muscles (flexor muscles) and triceps muscle (extensor muscle). Stretch receptors inside the muscles detect the state of contraction of the muscles. The receptors send impulses to the CNS along sensory neurones. When one muscle contracts sensory input is fed to the brain and also to reflexes that inhibit contraction of the other muscle.

Antagonistic action of the muscles that move the forearm is shown in Figure 30.

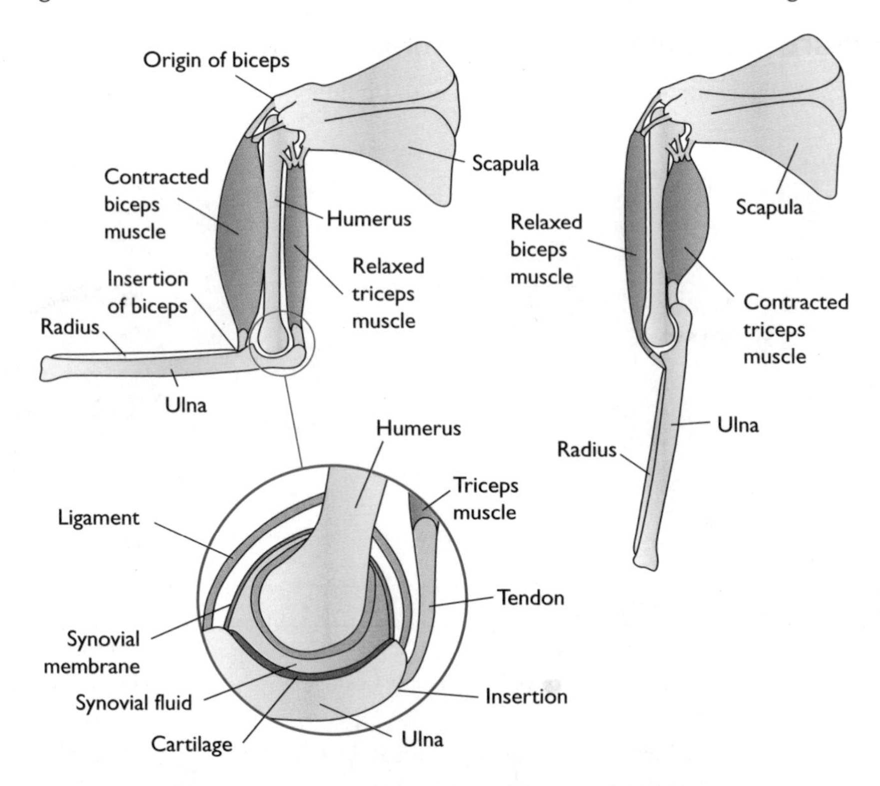

Figure 30 Antagonistic action of the muscles that move the forearm about the hinge joint at the elbow

Muscle contraction

There are animations of the process of muscle contraction at:

http://highered.mcgraw-hill.com/sites/0072495855/student_view0/chapter10/

Figure 31 shows what happens in a striated muscle fibre after it has been stimulated by a motor neurone.

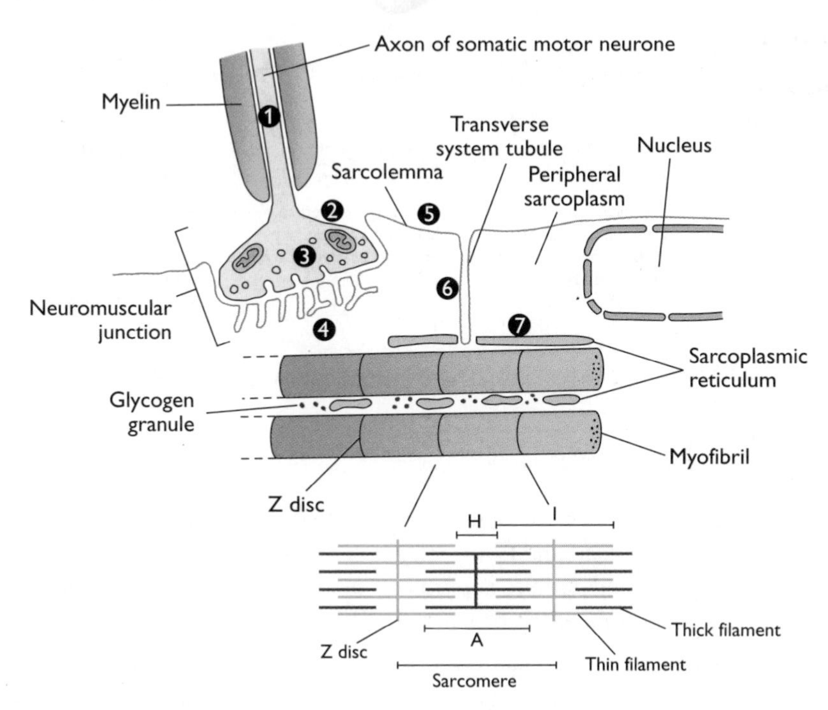

Figure 31 The events that occur in a striated muscle fibre following stimulation by a motor neurone. The numbers refer to the text

Motor neurones terminate at neuromuscular junctions. These are similar to synapses in that an impulse is transmitted across a gap. Action potentials travel down the motor neurone (1) and stimulate calcium ions to enter (2), causing vesicles of neurotransmitter to fuse with the membrane (3). Acetylcholine binds to receptors, causing an inflow of sodium ions (4) that depolarise the sarcolemma (an end-plate potential). If above threshold, action potentials are transmitted along the sarcolemma (5) and down into transverse system (T-system) tubules (6). These membranes have the same channel proteins and pumps that are found in neurones.

The impulse is coupled to the contraction mechanism. Depolarisation of T-system tubules causes calcium channels in the sarcoplasmic reticulum to open and release calcium ions into the sarcoplasm (7). Calcium ions act as second messengers to stimulate movement of the muscle myofibrils.

When calcium ions bind to troponin it causes a second protein, tropomyosin, to move, exposing binding sites for myosin on thin filaments. Myosin heads move towards the thin filament, bind to actin and swivel as is shown in Figure 32. This swivelling motion is the power stroke that moves the thin filaments closer together, reducing

Examiner tip

More detail about muscle contraction is given in Question 7 on pp. 100 and 102.

sarcomere length. The combined effect of shortening in all the sarcomeres shortens the length of the myofibril. The combined effect of shortening all the myofibrils in all the muscle cells shortens the length of the muscle. When there is no action potential in the sarcolemma, calcium is pumped back into the sarcoplasmic reticulum and contraction stops.

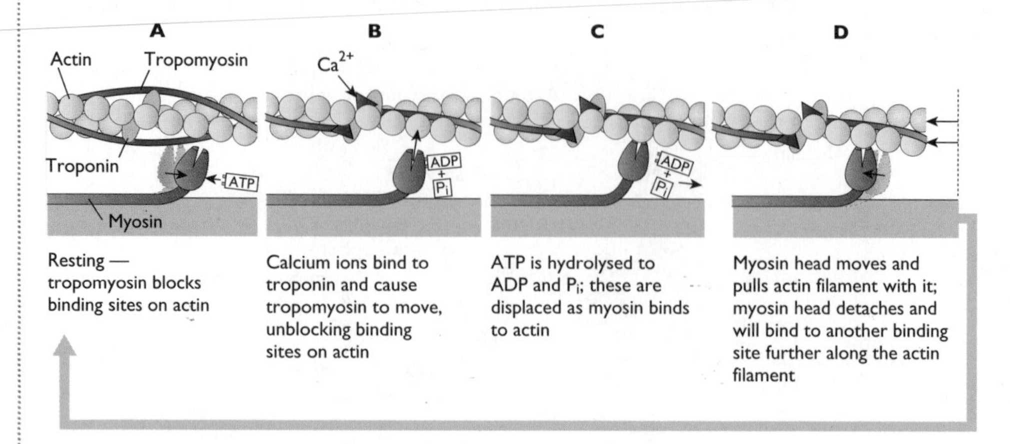

Figure 32 Sliding filaments: changes that occur in a myofibril during contraction (see the results in Question 7 on pp. 100 and 102)

The myosin head is an ATPase, which hydrolyses ATP to ADP and P_i. When it moves during the power stroke ADP and P_i are released. ATP then binds to the myosin head and causes the myosin and actin to separate. The myosin head returns to its original position and is ready to repeat the process.

There is little ATP available readily within muscle tissue. There is a store of creatine phosphate (CP) within the sarcoplasm where an enzyme transfers phosphate to ADP to resynthesise ATP. There is enough ATP and CP for about 6–8 seconds of exercise. Once this is exhausted, ATP must be synthesised from respiration. At the start of exercise or during short-term strenuous exercise, such as weight lifting and sprinting, the energy comes from anaerobic respiration of muscle glycogen with the production of lactate. This is unsustainable in the long term, so during aerobic exercise (e.g. distance running) aerobic respiration of carbohydrate and fat occurs.

'Fight or flight'

During a dangerous or stressful situation, the nervous and endocrine systems work together to coordinate responses from the whole body. Impulses from the sensory organs travel along sensory neurones in the somatic sensory system. These may stimulate reflexes through the spinal cord and brain. Impulses also travel to the cerebrum, which makes decisions about how to respond.

Impulses in somatic motor neurones travel to skeletal muscles to coordinate appropriate movements (stand and fight, or turn and flee). Impulses travel along sympathetic neurones in the cardiac accelerator nerves to increase the output of the heart by increasing the heart rate and stroke volume. Sympathetic innervation causes vasoconstriction in the gut and skin and vasodilation in muscles. Blood is distributed to where it is needed most to deliver oxygen and glucose. Some sympathetic pre-ganglionic neurones terminate in the adrenal glands, where they stimulate the secretion of adrenaline, which works alongside the sympathetic system and stimulates the liver to convert glycogen to glucose, increasing the blood glucose concentration.

Examiner tip

It is a good idea to make a table to compare the actions of the sympathetic nervous system with those of adrenaline. The two work together, but in different ways. See p. 28 of the Unit Guide for F214.

Synoptic links

Understanding the role of the nervous system in controlling movement builds on the work on structure and function of neurones (pp. 15–27 in the guide to Unit F214). The roles of aerobic and anaerobic respiration in muscle contraction depend on the biochemistry of these processes (pp. 60–71 in the guide to F214).

Animal behaviour

Key concepts you must understand

Animals show two main types of behaviour:

- innate (inborn) which is programmed genetically and is not learnt, although it might be modified by experience
- learned behaviour that is acquired during the lifetime of the individual

Examiner tip

Read examples 32 and 33 in the Unit Guide for F213/F216 to find out about practical investigation of taxes and kineses. These could be contexts for questions in the exam for F215.

Key facts you must know

Innate behaviour

Table 17 Innate behaviour

Type of innate behaviour	Stimulus	Response	Examples
Simple reflex	Stretch of tendons/muscles, vibrations	Fast, automatic and does not need any conscious thought	Knee-jerk reflex in humans; escape reflex of earthworms
Taxis (plural: taxes)	Directional — light, chemical	Directional movement towards or away from the source of the stimulus	Negative phototaxis of maggots
Kinesis (plural: kineses)	Humidity, light, temperature	Rate of movement/turning is related to the intensity of the stimulus, *not* its direction	Woodlice: faster movement in dry/bright conditions; slower in the damp and dark

Examiner tip

When reading about the different types of behaviour it is a good idea to look at short films showing the behaviour described — there are plenty on YouTube. Reading about the work of the researchers named below (p. 70) will also help you with the *How Science Works* aspect of this topic.

The advantages of innate behaviour are that it is there right at the start of an animal's life and has survival value. Behaviour patterns are the result of natural selection, so they adapt young animals to survive.

Learned behaviour

- **Habituation** — if a stimulus is repeated and it is harmless, an animal learns to ignore it. Opening an umbrella in front of a young calf will cause it to flinch and move away. After a while, it ignores the stimulus.

- **Imprinting** — during a sensitive period in an organism's life (usually after birth or hatching) the animal learns the features of its parent. This sensitive period does not last long (Konrad Lorenz).
- **Classical conditioning** — an innate response, such as a simple reflex, is modified. The example of Ivan Pavlov's dogs is shown in Table 18.

Table 18 Classical conditioning

Stimulus	Role of nervous system	Response
Unconditioned stimulus: sight or smell of food	Somatic sensory neurones from nose/eyes to cerebrum linked to autonomic (motor) neurones to salivary glands	Unconditioned response: salivation — production of saliva by salivary glands
Conditioned stimulus: sound of bell ringing	Somatic sensory neurones from ear to association centre and via motor neurones to salivary glands	Conditioned response: salivation

- **Operant conditioning** — an animal carries out an action so as to receive a reward or to avoid an unpleasant experience such as a mild electric shock. This can be demonstrated using Skinner boxes in which rats or pigeons move around, accidentally press a lever and receive some food. The animals learn to associate an operation (pressing a lever) with a reward (B. F. Skinner).
- **Latent learning** — animals explore and learn details of their surroundings. This is useful information that they may need later (Edward Tolman and C. H. Honzik).
- **Insight learning** – animals integrate past experiences to show a new piece of behaviour. Wolfgang Köhler studied chimpanzees in captivity in the Canary Islands. He gave them boxes to play with. They used them to reach bananas. When bananas were hung a long way above them, they stacked up the boxes to reach them.

Primates, such as chimpanzees, gorillas and orang-utans, show complex social behaviour. This is seen, for example, when a group of chimpanzees is approached by a leopard, which is one of their few predators. Chimpanzees on their own, especially young animals, may be quite defenceless against such a predator. However, there is strength in numbers and older animals in the group work together to scare away the leopard by screaming, tearing up small trees and using them to threaten the predator. For this to work effectively, the animals must be able to communicate with each other. They use signals such as noises — grunts, hoots, screeches and whimpers — as well as a variety of gestures and facial expressions. Mutual grooming reinforces the bonds within such a group and there are also behaviours that maintain a hierarchical order with an alpha male at its head. All this is related to the lengthy period of development of the young, which requires parental care. It is worth reading work on chimpanzees (Jane Goodall), gorillas (George B. Schaller and Dian Fossey) and orang-utans (Biruté Galdikas).

Dopamine receptors

In the brain there are many different neurotransmitters (e.g. acetylcholine, GABA and dopamine) that modulate the behaviour of neurones. Dopamine is involved with voluntary movements. People who suffer from Parkinson's disease have reduced levels of dopamine and find it difficult to make such movements. On the surface of postsynaptic neurones there are five different receptors for dopamine (D1, D2, etc.) including the D4 subtype which has its gene locus (***DRD4***) on chromosome 11.

Examiner tip

Make sure you learn the differences between these two types of conditioning as it is easy to confuse them. 'Pavlov's dogs' is an example of *classical* conditioning.

Examiner tip

Make a list of these different forms of learning and find examples. For each example, learn the name of the animal, the stimulus to which it responds and describe the response. For example, earthworms respond to vibrations in the ground or passing shadows by retreating into their burrows.

Knowledge check 34

Identify the type of behaviour shown by an earthworm when it retreats into its burrow in response to a passing shadow.

Knowledge check 35

Distinguish between innate and learned behaviour.

This is a G-protein-coupled receptor that inhibits adenyl cyclase. Mutations in the receptor protein are linked with disorders of the autonomic nervous system as well as hyperactivity, novelty seeking, alcoholism, drug abuse and schizophrenia. These mutations involve repeats of a 48-base-pair section in the gene. These repeats lead to repeats in the primary structure of the receptor and change its affinity for dopamine. Knowledge of the shape of the variants of D4 and other receptors can help doctors prescribe drugs appropriate for each individual. This is one of the advantages of sequencing genes (see p. 44) and having knowledge of the tertiary and quaternary structures of proteins.

Knowledge check 36

The gene for the dopamine receptor D4 has many alleles, which have different numbers of base pairs. Explain why any one person may have no more than two of these alleles.

Synoptic links

D4 is a good topic for testing your understanding of many themes in this unit and in Unit F214. The following could be asked in this context:

- gene sequencing and ***DRD4*** sequences
- shape of receptors and cell signalling involving dopamine
- neurotransmitters and drugs that interact with their receptors
- voluntary movement
- role of D4 in inhibiting adenyl cyclase
- cAMP as a second messenger

You may be asked to compare neuromuscular junctions and interneuronal synapses – compare Figure 31 on p. 67 with a diagram of a synapse (p. 25 of the guide to Unit F214).

Summary

- Plants respond to changes in their environments to avoid predation and abiotic stress (e.g. the effects of drought).
- A tropism is a plant growth response to stimuli such as gravity and the direction of light.
- Plant hormones coordinate activities in plants. Auxins coordinate tropisms and apical dominance; gibberellins control stem elongation. Various hormones coordinate leaf fall in deciduous plants.
- Auxins are used commercially as weedkillers and rooting compounds; gibberellins are used to promote growth of grapes and hasten the germination of barley.
- Animals respond to changes in their environments for many reasons, such as the search for food, the avoidance of predators and the location of mates and nesting sites.
- The human nervous system is divided into the central nervous system (brain and spinal cord) and the peripheral systems (cranial and spinal nerves).
- The autonomic nervous system is composed of motor neurones that control activities over which we do not have much voluntary control. The parasympathetic system controls conservation of body resources and the sympathetic system works, along with adrenaline, to coordinate responses to danger, stress and physical activity.
- The main regions of the human brain are cerebrum, cerebellum, medulla oblongata and hypothalamus.
- Innate behaviour is inborn and does not need to be learnt. The advantage for many organisms is that it happens straight away when a young animal may not have any protection from an adult.
- Escape reflexes, taxes and kineses are examples of genetically determined innate behaviours.
- Habituation, imprinting, classical and operant conditioning, and latent and insight learning are all examples of behaviours that are learned by animals.
- Primates show many forms of complex behaviour within family or larger groups. These are forms of social behaviour, which help to secure food sources, provide protection and assist the rearing of young animals.
- Understanding the roles of the dopamine receptor, D4, in synapses in the CNS may help explain various forms of human behaviour.

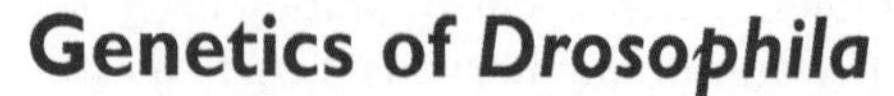

Genetics of *Drosophila*

Almost all the genetics in this guide is based on the fruit fly *Drosophila melanogaster*. This animal has been used in studies on genetics for about 100 years. There are computer simulations of genetic crosses with fruit flies that you could use if your school or college does not have the resources for you to carry out crosses with real flies. Before you read the sections in this guide about fruit flies, look at photographs and drawings of male and female fruit flies, read about the life cycle and look at a map of the chromosomes to locate the genes mentioned in this guide.

For the chromosome map see:

http://bio3400.nicerweb.com/Locked/media/ch05/05_14-Drosophila_map.jpg

For phenotypes see:

http://cgslab.com/phenotypes/

For help with many aspects in Modules 1–3 see:

http://learn.genetics.utah.edu/

To help your understanding of genetics, download a demonstration version of the program FlyLab at:

www.biologylabsonline.com

This allows you to simulate a number of different crosses to illustrate all the aspects of genetics in the specification using a large number of genes in *Drosophila*. It also allows you to do chi-squared tests on the data it generates. Beware — the free trial lasts one day and it will take you some time to learn how to use the program and read the information provided.

Questions & Answers

The unit test

The examination paper will be printed in a booklet, in which you will write all your answers. The paper will have about seven or eight questions, each divided into parts. These parts comprise several short-answer questions (no more than 3 marks each) and questions worth 4, 5 or 6 marks. There will also be some part-questions requiring extended answers, for up to 10 marks each. These are likely to be the questions where your written communication skills are assessed. You can expect questions to cover more than one module of the unit, as in Questions 1 and 3 in this guide, which examine some of the content of Modules 1 and 2. You can also expect questions that cover topics in the other modules (AS and A2); these are the synoptic questions. The unit test is worth 100 marks and lasts 2 hours.

As you read through this section, you will discover that student A gains full marks for all the questions. This is so that you can see what high-grade answers look like. Notice how student A uses technical terms from the specification to good effect. Remember that the minimum for grade A is about 80% of the maximum mark (in this case around 80 marks). Student B makes a lot of mistakes — often these are ones that examiners encounter frequently. I will tell you how many marks student B gets for each question. If the overall mark for the paper is about 40% of the total (around 40 marks), then the student will have passed at grade E standard. Use these benchmarks when trying the questions yourself.

The quality of your written communication is assessed in two questions; these questions are indicated on the exam paper. You could be asked to describe the steps of a process in the correct sequence, to compare and contrast features, or relate structure to function. There are plenty of potential examples in the Content Guidance section of this guide. You are expected to use technical terms correctly, especially in the longer-answer questions.

On p. 104 there is a summary of the mistakes made by student B. This should help you to identify what the student should have done during revision and in the examination in order to gain a better mark.

Examiner's comments

Examiner's comments on the questions are preceded by the icon e. They offer tips on what you need to do to gain full marks. Examiner's comments on the students' answers are preceded by the icon e and indicate where credit is due. In student B's weaker answers they also point out areas for improvement, specific problems and common errors, such as lack of clarity, weak or non-existent development, irrelevance, misspellings, misinterpretation of the question and mistaken meanings of terms.

Question 1 **Gene control**

The bacterium *Escherichia coli* synthesises the enzymes of glycolysis (for the metabolism of glucose) all the time. The enzyme β-galactosidase catalyses the hydrolysis of the disaccharide lactose into glucose and galactose. This enzyme is only synthesised when lactose is present. The synthesis of β-galactosidase is controlled by a system known as the *lac* operon.

(a) Explain what is meant by the term *operon*. (2 marks)

e Here you need to remember a suitable definition. This is why it is a good idea to write your own glossary to help your revision.

(b) Explain the advantage of synthesising β-galactosidase only when lactose is present. (2 marks)

e In this question, think of the advantages to the bacteria. Do not just think about the operon.

Using a sterile pipette, a sample was taken from a laboratory culture of *E. coli*. The sample was transferred to a large volume of sterile culture medium containing 0.001 mol dm^{-3} of glucose and 0.001 mol dm^{-3} of lactose.

The population of live bacteria in the culture was determined at 15-minute intervals. The results are shown in Figure 1.

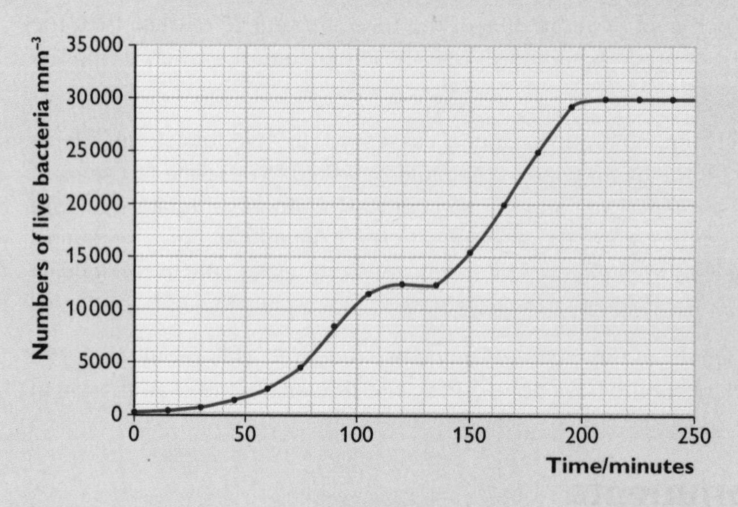

Figure 1

(c) Use the information in Figure 1 to answer the following questions.

(i) Describe and explain what happens to the population of *E. coli* during the first 100 minutes. (4 marks)

e There are two command words in this question. Make sure you answer both, giving at least two different points for each as there are 4 marks available.

(ii) Explain why the population remained stable after 120 minutes and then increased after 135 minutes. (4 marks)

e There is no need to give any data in answer to this question. Think back to the operon and its control.

(iii) Suggest why the number of live bacteria was counted rather than the total number of bacteria in each sample. (1 mark)

Total: 13 marks

e This is a 'suggest' question, so do not think you know the answer. You have to work it out. The clue is in the question.

Student A

(a) An operon is an arrangement of genes with common regulator sections of DNA (promoter and operator regions). The structural genes in the *lac* operon are close together and are transcribed together. The regulator sections determine whether the genes are transcribed or not. Transcription factors decide whether the genes are switched 'on' or 'off'.

Student B

(a) An operon is a regulator gene and some genes that code for enzymes.

e Student A gives a thorough answer generalising from the *lac* operon to the other operons. It is worth finding information about the *trp* operon in *E. coli*, which is involved with the synthesis of the amino acid tryptophan. The control is slightly different from that of the *lac* operon. Student B has not developed the answer to explain that operons determine whether or not genes are transcribed by RNA polymerase. There is no reference to operator and promoter regions. Student B fails to score.

Student A

(b) It is a waste of energy in the form of ATP to synthesise enzymes unless the correct conditions (e.g. food) are available. It is wasteful to make enzymes when they are not wanted. Some enzymes are only needed at certain times during the life of a cell.

Student B

(b) Lactose is not available, so there is no need for that enzyme. The repressor substance binds to the operator region so that the gene for the enzyme cannot be made.

e Student A gives the reason why it is an advantage and also refers to synoptic material by mentioning ATP from Unit F214. Student B has not developed the idea that there is no need for the enzyme in terms of not wasting energy or materials, such as the amino acids that are needed to make the β-galactosidase. Student B has written 'so that the gene for the enzyme cannot be

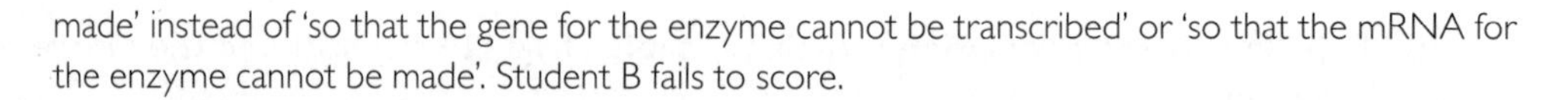

made' instead of 'so that the gene for the enzyme cannot be transcribed' or 'so that the mRNA for the enzyme cannot be made'. Student B fails to score.

Student A

(c) (i) When the bacteria are first put into the environment they are in a lag stage. During the first 45 minutes the bacteria become used to the environment and its nutrient supply. The cells make enzymes and transport proteins for membranes. Next is the log stage — this is between 45 and 100 minutes when the bacteria multiply rapidly and the population reaches 10 000 per mm^3. Then the increase starts to slow as environmental resistance sets in.

Student B

(c) (i) The population shows sigmoid growth during this time. To begin with the bacteria do not divide at all as they are not used to the environment. Then there is a dramatic increase in numbers because there are no limiting factors.

e When a question starts 'Using the data...' or 'With reference to Figure...' this means that you should quote some of the data from the graph, table or other material that is provided. Student B has not done this so has not gained an easy mark. It is possible to gain full marks for 'describe' questions without quoting data, but it is more difficult. Here, student B refers to sigmoid growth and gains a mark for using that term correctly. However, another mark could be gained by quoting some figures, as student A has done. 'To begin with' and 'dramatic increase' are not precise enough to gain marks for description; 'no limiting factors' should be illustrated by at least one example. Student B gains 1 mark.

Student A

(c) (ii) The population is stable because the glucose is now exhausted and the rate of production of bacteria by binary fission is equal to the death rate. There is a food supply available (lactose), but the bacteria do not have the lactase enzyme. Some lactose enters the bacteria, binds to the repressor substance removing the inhibition so the gene for lactase is transcribed and translated. This didn't happen earlier because the *lac* operon is turned off when glucose is present because it is easier to respire glucose as there is no need for an enzyme to hydrolyse glycosidic bonds.

Student B

(c) (ii) This stage is called the 'stationary phase'. Here the rate of growth is constant because the nutrients have run out. The number of bacteria made is equal to the number that dies. Excretory products will also inhibit the metabolism of the bacteria.

e Prompted by the name of the enzyme, student A refers to the glycosidic bond in disaccharides. This is good use of material from an AS unit. Student B does not respond to the second part of the question that asks for an explanation of the second part of the graph. Always look out for two-part questions like this. Examiners may ask for a description *and* an explanation, or advantages *and* disadvantages, and embolden the 'and' so you don't miss the instruction as Student B has done here. Student B gains 2 marks.

Student A

(c) (iii) Some of the bacteria would be dead but not decomposed so you would see them if you put a sample under the microscope.

Student B

(c) (iii) Many would be dead so you would not see the plateau when you plot the total numbers as bacteria are still dividing during the stationary phase.

(e) Both students have the right idea here. Student B has 1 mark. When investigating the growth of a population of microorganisms it is important to count live organisms. See p. 39 to remind yourself how this is done.

(e) **Student B gains 4 marks out of 13 for Question 1.**

Question 2 **Gene interaction**

The synthesis of the blue pigment, malvidin, in flowers of *Primula* is controlled by two unlinked genes, K/k and D/d. The dominant allele, K, codes for the production of malvidin. No pigment is produced by plants with the genotype kk and their flowers are white. The production of the pigment is suppressed by the dominant allele D, but not by the recessive allele, d.

(a) State the genotypes of *Primula* plants with blue flowers. (1 mark)

e Note that there are two gene loci in the information above; both must be included in your answer. See p. 21 for advice on writing genotypes in dihybrid crosses. Always read the whole question before answering the first part. You can see how other genotypes are written in (b).

(b) Two *Primula* plants with the genotypes KkDd and kkdd were crossed. Draw a genetic diagram to show the phenotypes of the parent plants, the gametes and the genotypes and phenotypes of the offspring. Give the ratio of phenotypes in the offspring. (5 marks)

e You might be given a grid with subheadings (parental phenotypes, parental genotypes etc.) to complete when answering this type of question. If not, follow the pattern in the genetic diagrams used in this guide between pages 20 and 25.

(c) (i) State the name given to the gene interaction shown in the control of flower colour in *Primula*. (1 mark)

(ii) Suggest how allele D interacts with the locus K/k to give the results you have shown in part (b). (2 marks)

e You should have identified the type of interaction while reading the introduction to this question. You should write this down on the exam paper as you read. It will help you to focus on the appropriate learning outcomes from the specification.

Some students investigated the inheritance of two genes in the fruit fly, *Drosophila melanogaster*. Pure-bred flies with straight wings and dark red eyes were crossed with pure-bred flies with curved wings and brown eyes. All the F_1 generation had straight wings and dark red eyes.

In a test cross, the females of the F_1 generation were crossed with males with curved wings and brown eyes. The students expected a ratio of 1:1:1:1 in the test cross offspring.

The results were as follows:

• **straight wings and dark red eyes**	**79**
• **curved wings and dark red eyes**	**25**
• **straight wings and brown eyes**	**23**
• **curved wings and brown eyes**	**73**
Total	**200**

A chi-squared (χ^2) test was carried out to see if the number of each phenotype was in agreement with a 1:1:1:1 ratio. The value of χ^2 is 54.48.

Table 1 shows critical values of χ^2 at different levels of significance and degrees of freedom.

(d) (i) Using Table 1, state the conclusion that would be drawn from the results of this investigation and explain how you reached this conclusion. (3 marks)

ⓔ There are some key words you should use in answer to this type of question. Make a note of these in the margin of the exam paper and make sure they are in your answer.

(ii) Explain why the results are different from the expected results. (3 marks)

ⓔ This is where you need to apply your knowledge of inheritance to a specific example that you will not have seen before.

Table 1 Distribution of χ^2

Degrees of freedom	Probability, *p*				
	0.10	**0.05**	**0.02**	**0.01**	**0.001**
1	2.71	3.84	5.41	6.64	10.83
2	4.61	5.99	7.82	9.21	13.82
3	6.25	7.82	9.84	11.35	16.27
4	7.78	9.49	11.67	13.28	18.47

(e) Milk yield in cattle is influenced by both genetic and environmental factors. Explain why it is necessary to consider environmental factors when using selective breeding to improve milk yield. (3 marks)

Total: 18 marks

ⓔ Think about the different environmental factors that are likely to influence milk yield. These should help you to frame your answer.

Student A

(a) KKdd, Kkdd

Student B

(a) K_dd

ⓔ Student A has given the only two genotypes that result in flowers containing malvidin. To produce this pigment there must be a **K** allele and no **D**. It is perfectly acceptable to use the dash (_) in genotypes. You can write this on the exam paper when working out genetics problems. However, this question asks for the *genotypes* (plural) and so student B should have realised that more than one genotype is required. If there had been 2 marks available, then the examiner might have awarded 1 mark. In this case, no mark is awarded.

Student A

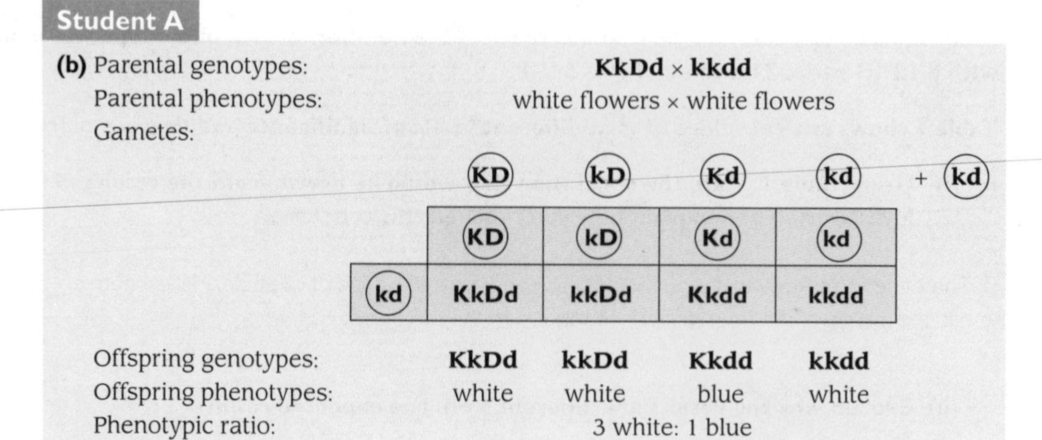

(b) Parental genotypes: **KkDd × kkdd**

Parental phenotypes: white flowers × white flowers

Gametes: (KD) (kD) (Kd) (kd) + (kd)

	(KD)	(kD)	(Kd)	(kd)
(kd)	**KkDd**	**kkDd**	**Kkdd**	**kkdd**

Offspring genotypes: **KkDd** **kkDd** **Kkdd** **kkdd**

Offspring phenotypes: white white blue white

Phenotypic ratio: 3 white: 1 blue

Student B

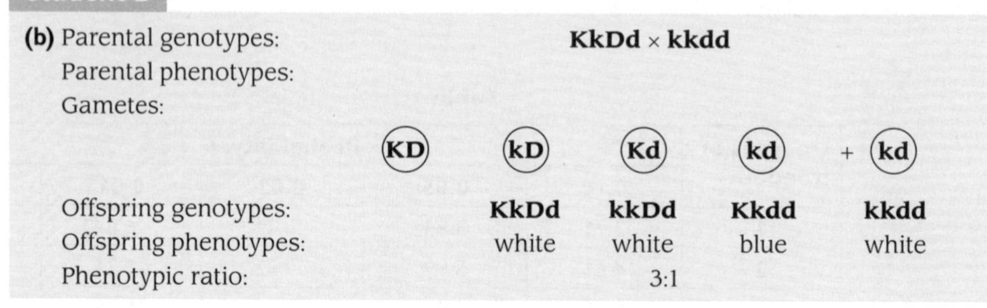

(b) Parental genotypes: **KkDd × kkdd**

Parental phenotypes:

Gametes: (KD) (kD) (Kd) (kd) + (kd)

Offspring genotypes: **KkDd** **kkDd** **Kkdd** **kkdd**

Offspring phenotypes: white white blue white

Phenotypic ratio: 3:1

e Student A has completed all the sections of the genetic diagram. The side headings are usually provided on the examination paper to help you with the steps of the genetic diagram. Student B has neglected to give the parental phenotypes. Student A has given a Punnett square to show all the fusions that take place. This is good practice. Although student B has not made a mistake in working out the genotypes of the offspring, it is easy to do so if you do not use a Punnett square. Both students have written the phenotype underneath each genotype, which is essential to gain a mark, but student B has not written in the flower colours in the phenotypic ratio, so loses a mark. In this question the flowers without malvidin are described as white. In fact, the colour of the flowers depends on other genes. Control of flower colour in plants and coat colour in mammals are examples of epistasis that you could be asked about in the examination. Student B gains 3 marks.

Student A

(c) (i) Dominant epistasis

(ii) The dominant allele could code for an inhibitor protein that stops transcription of the gene for the enzyme that makes malvidin. It does this by binding to the operator region. Alternatively, the inhibitor might bind to the enzyme, changing the shape of its active site so that it doesn't work.

Student B

(c) **(i)** Epistasis

(ii) The gene could code for an enzyme that breaks down the malvidin once it is made. Or it could inhibit malvidin.

ⓔ Both students gain the mark for **(c) (i)**. If there had been 2 marks for this part-question then it would have been necessary to refer to dominant or recessive epistasis. In **(c) (ii)**, student B should have said 'The dominant allele…' rather than 'The gene…' but in this case it is clear what is meant and the student gains 1 mark. 'It could inhibit malvidin' does not tell us much. If you say that something is an inhibitor you should always explain or suggest how it works. Another explanation is that allele **D** codes for an enzyme that breaks down malvidin.

Epistasis is a difficult topic — especially predicting the phenotypic ratios without first drawing a genetic diagram. It is likely that the ratios you are asked to predict are those obtained when crossing organisms that are heterozygous for the two interacting genes. This always gives a 4 × 4 Punnett square with all the possible combinations of dominant and recessive alleles of the two loci. A test cross, such as that shown in **(b)**, reveals the different types of gene interaction without your having to work out the phenotypes of 16 genotypes. You can find good explanations of the common epistatic ratios seen in the F_2 generation at: **www.ndsu.nodak.edu/instruct/mcclean/plsc431/mendel/mendel6.htm**

Student A

(d) **(i)** There is a significant difference between the observed results and the results expected by independent assortment of the alleles of the two genes (1:1:1:1). This means that the null hypothesis is rejected. This is because the value for χ^2 is greater than the critical value of 7.82 at the 5% significance level. It is greater than 16.27, so we can be 99.9% certain that the results are not due to chance.

Student B

(d) **(i)** The degrees of freedom = 3.
The critical value for χ^2 at $p = 0.05 = 7.82$. As the value of 54.48 is greater than the critical value, the null hypothesis can be rejected and, therefore, these results were not obtained simply by chance.

ⓔ Both students gain 3 marks here. 'The conclusion' mentioned in the question means that you must say whether the null hypothesis is accepted or rejected. It is always a good idea to include the level of significance, as student A has done.

Student A

(d) **(ii)** The two genes are linked on the same chromosome so independent assortment did not happen. If it had done, then the four phenotypic classes would be present in approximately equal numbers and the value for χ^2 would be less than the critical value. The recombinant classes were obtained by crossing over. Chiasmata formed between non-sister chromatids and there was exchange between them.

Student B

(d) (ii) The expected results from the test cross are in the ratio 1:1:1:1, so this would be 50:50:50:50 for the different phenotypes. The results could not have been obtained by chance — for example by the random fusion of gametes. The chi-squared test shows that there must be another explanation.

ⓔ Student A realises that the results are obtained by crossing over between the two gene loci that are linked on the same autosome. Student B has simply repeated the explanation given in part **(i)**, so fails to score.

Student A

(e) Milk yield is a characteristic that shows continuous variation and is therefore influenced by both genotype and the environment. It is a polygenic feature as it is controlled by many genes. The production of sugars, proteins and fats in the milk is also influenced by the quality and quantity of food given to the cows. When breeders choose females to use in breeding programmes they must make sure that they all have the same feed. If all the cows are kept under the same conditions any differences between them will be due to the alleles of the genes that control milk yield. They can then choose the best milkers to be inseminated by the best bulls.

Student B

(e) Selective breeding means taking certain males and females and crossing them. Milk yield is a sex-limited feature only expressed in females. The males are progeny tested to find those that have daughters which give plenty of milk. Females with high yield are given good conditions, so the only factor controlling their milk yield is the environment. The best females (the ones which produce most milk) have the 'best' genes and should be used.

ⓔ Student B has not really answered the question. The answer is a description of selective breeding. There is almost a mark for writing about 'good conditions' but since milk yield is an example of selective breeding given in the specification the examiner would expect a more precise answer. Student A mentions the quality and quantity of food. This answer could have explained more about the diet, but the important point is that the environment is the same for all of the cows so the genetic component can be assessed. Student B refers to the 'best' genes. This is not a good idea even though the student has used inverted commas to show that is not exactly what is meant. All cows have the same genes. It is the alleles of those genes that differ, with some alleles contributing more to the milk yield than others. Student B fails to score.

ⓔ **Student B gains 8 marks out of 18 for Question 2.**

Question 3 **Genomes, gene sequences and selection**

The gene for the β-polypeptide of human haemoglobin was isolated from chromosome 11. The chain termination method using the four dideoxynucleoside triphosphates (ddNTPs) was used to sequence the gene. The four ddNTPs were each labelled with a different fluorescent dye.

Figure 2 shows the results of detecting the four different ddNTPs for the first 12 nucleotides of the gene.

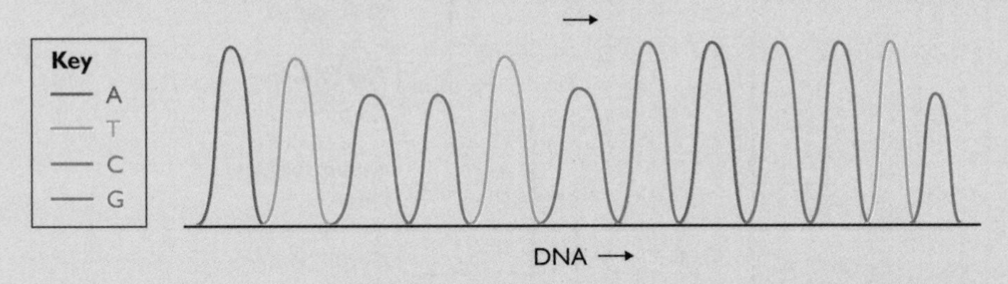

Figure 2

(a) Using the information in Figure 2, write out the sequence of bases for the gene. Write one letter in each box. (2 marks)

1	2	3	4	5	6	7	8	9	10	11	12

ⓔ This is an easy start to the question. Do not rush a question like this as you are likely to make a mistake and lose these easy marks.

(b) The sequencing process began with a length of DNA isolated from chromosome 11. Outline the steps involved in sequencing this gene to obtain the results shown in Figure 2. (5 marks)

ⓔ You could answer this question with numbered points. They might help you to get the sequence correct. If you make a mistake it is easy to change the number rather than having to cross out and rewrite.

(c) State what is meant by the term *biological species*. (1 mark)

ⓔ Here is another example that shows you the benefits of writing your own glossary.

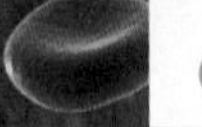

Genes that code for the β-polypeptide of haemoglobin of seven different species of primate were sequenced. Figure 3 below is based on the data obtained.

(d) With reference to Figure 2 and Figure 3, explain how data from gene sequencing can be used to assess the evolutionary relationships between different species. (3 marks)

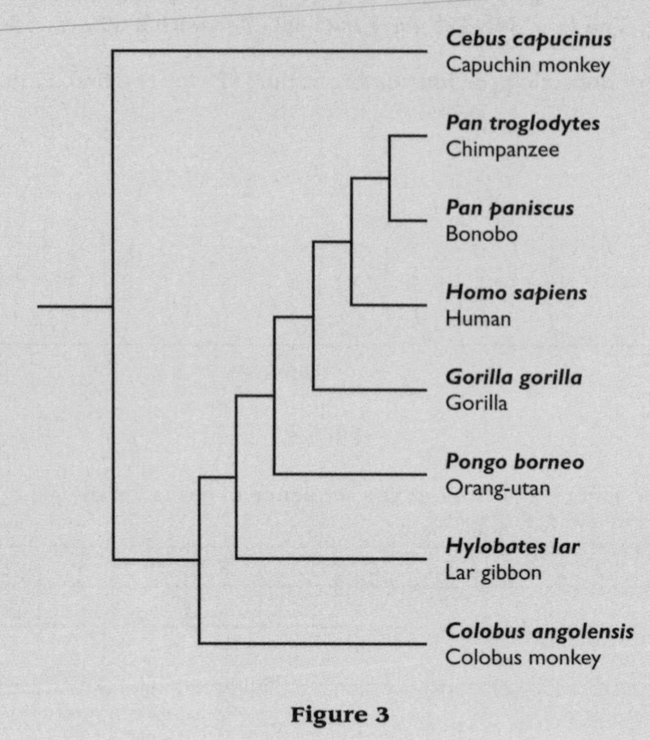

Figure 3

'With reference to…' means you can give information from the figures in your answer. These are usually easy marks, so spend some time looking carefully at both figures.

Cepaea nemoralis **is a species of snail found widely in the UK. The presence or absence of bands on the shell is controlled by a single gene (B/b). The unbanded allele (B) is dominant to the allele for banded (b). In a population of these snails, 36% showed the banded phenotype. In parts of the UK, the proportion of snails with bands in populations of *C. nemoralis* has remained constant for many years.**

(e) Use the Hardy–Weinberg principle to calculate the frequency of the allele for unbanded in this population. Explain how you arrive at your answer. (3 marks)

In some calculations, you may be given full marks even if you do not include any working. Here you must show your working. Write out an explanation *in words*. Do not just write down the numbers. You could include one or both equations (see p. 33).

(f) Explain how environmental factors act as stabilising forces to maintain allele frequencies from generation to generation. (3 marks)

Total: 17 marks

ⓔ This does not ask for an example, but you could use one in your answer. You might also find it useful to include a sketch graph of the variation in a feature that shows continuous variation, such as birth mass or wing length, (see p. 29) to help your explanation. The pattern (probably a bell-shaped curve) is the same from one generation to the next in this form of selection.

Student A

(a) ATGGTGCACCTG

Student B

(a) ATGGTGCAGGTG

ⓔ Student B has misread two of the bases from the diagram so fails to score. If 2 marks had been available, it is likely that the examiner would have awarded 1 mark. Look out for questions similar to this where information has to be translated from one form to another. These are relatively straightforward and are often included to give an easy start to a question.

Student A

(b) 1 The DNA is denatured by heating. This splits the hydrogen bonds and gives two polynucleotide chains.

2 The chain to be sequenced is amplified by PCR.

3 The lengths of DNA are placed into tubes along with a primer (a piece of DNA that is known to be complementary to part of the DNA being sequenced), four dNTPs (with A, T, C and G), DNA polymerase and one of four ddNTPs (e.g. ddATP) that terminate the replication.

4 After a while, the DNA fragments from the separate reaction mixtures are mixed together.

5 The fragments are separated using electrophoresis in a narrow tube.

6 A laser detects the fluorescent nucleotides as they pass along. The smallest fragments from the beginning of the sequencing by the polymerase will come out first. In this example, ddATP would come out first.

Student B

(b) The DNA is cut by restriction enzymes. There is a restriction site for each restriction enzyme. These cut the DNA into different-sized fragments. They are sorted by electrophoresis. The small fragments travel further towards the anode than the larger fragments. It is possible to see the fragments of DNA by pressing a nylon membrane on the gel. Some of the gel transfers to the membrane and a radioactive DNA probe is applied to the membrane where it binds to target sequences that have a complementary set of bases. This membrane is placed next to an X-ray film. The DNA which is now radioactive produces a 'fogging' on the film and you can read off the sequence.

ⓔ If you are asked to describe a process, it is perfectly acceptable to do this by writing numbered points, but you must get the steps in the right order. Numbered points are better than bullet points because, if necessary, you can refer to them by number in your answer. Bullet points are acceptable, but if you use them, make each bullet point a complete sentence unless it is a short list within a longer answer. Student A has given a thorough answer, starting with double-stranded DNA. Student B has remembered a different procedure — the one for cutting DNA into fragments using restriction enzymes and then using a probe to find a particular sequence. This would have occurred at an *earlier* stage in the gene sequencing procedure when researchers are cutting up much longer stretches of DNA containing many genes and non-coding regions. You cannot use this method to 'read off a sequence'. Student B has some correct information about electrophoresis and gains 1 mark. The information about Southern blotting using the nylon membrane is correct and is something you might have read about. Here it is not relevant.

Student A

(c) A biospecies is defined as all the individuals that interbreed to produce fertile offspring and are reproductively isolated from other species.

Student B

(c) A biospecies consists of all the organisms that can breed together to produce offspring that are fertile and themselves can interbreed.

ⓔ Both students have given enough information to gain the mark. The next question relies on knowledge of the species concept.

Student A

(d) The branching points in the diagram indicate where there are significant differences between the gene sequences. The polypeptides of haemoglobin function to transport oxygen, but there are differences in the base sequences that code for the order of the amino acids and also in the non-coding regions (e.g. introns — non-coding regions within a gene). Substitution mutations occur so that the sequences are different. These may either code for different amino acids or be silent mutations (often the third base is the one that changes and the new triplet codes for the same amino acid as the original). These will be neutral mutations as all organisms have functional haemoglobin.

So where the branching point is to the right the species are closely related as they have most of the base sequence in common (e.g. chimpanzee and bonobo). Where the branching point is to the left, the species are less closely related as there are more differences in the gene sequence (e.g. *Gorilla* and *Cebus*). This information is compared with information about other genes, non-coding regions of DNA and protein sequences to work out the evolutionary history (phylogeny) of the primates.

Student B

(d) This shows that humans and chimps are more closely related to each other than either is related to the other primates, e.g. gorilla. The monkeys are not closely related at all. The differences between gene sequences show that mutation must have occurred but the protein that the gene codes for still works as an oxygen carrier.

e This phylogenetic diagram has been drawn using limited data from one gene. However, looking at the base sequences of individual genes usually confirms the relationships between species that have been discovered from other sources of information such as morphology, anatomy, physiology and protein sequencing. Student B describes what the diagram shows, but does not explain how the gene sequences are used and so fails to score. Student A has written far too much for a 3-mark question. There is a danger if you do this that you will run out of time. The detail has been included here to help you understand this topic, which is identified in the specification as an aspect of *How Science Works*, so it would be a good idea to read more about it. Note that the biospecies concept is difficult to apply. The phylogenetic species concept uses information that we can observe and which can be used to explore the evolutionary history of species or higher taxa, such as the primates.

Student A

(e) Banded is the recessive phenotype, so 36% are homozygous recessive
36% = 0.36
$\sqrt{0.36} = 0.6 = q$ (frequency of the recessive allele)
p = frequency of the dominant allele
$p = 1 - 0.6$
$p = 0.4$

Student B

(e) If 36% are banded, then 64% are unbanded.
Unbanded is dominant, so unbanded snails are either homozygous dominant or heterozygous.
$p^2 + 2pq + q^2 = 1$
We know that $q^2 = 0.36$, so q must be 0.6.
Therefore, as $p + q = 1$; $p = 1 - q$; q must be 0.4.
The frequency of the dominant allele is 0.4 or 40%.

e Both students have the right answer and have given a suitable explanation. Note that this question is not like the usual calculation questions that have 2 marks. Here, the examiner asks for an explanation, rather than giving the usual instruction to 'show your working'. It might help the answer to use symbols for the alleles, although it is not necessary. Student B's method is rather long-winded, but work through it if you are unsure about how to answer questions on the Hardy–Weinberg principle. The question says that these frequencies have existed for a long time — you can see that the population is in equilibrium because you can calculate the frequencies in the next generation and they are the same as in the existing one. Student B gains 3 marks.

Student A

(f) Abiotic factors, such as temperature, climate, availability of shelter, and biotic factors, such as food availability, predators and disease, act as agents of selection on populations. If the environment is stable, then the factors that act are the same year after year. Organisms that are not well adapted do not survive to breed and pass on their alleles. In the snails if there are two alleles for a gene and both exist in large numbers in the gene pool, there must be advantages to keeping both.

Student B

(f) Banded and unbanded snails are present in the population in high percentages (36% and 64%). If these have existed for a long time, then they survive, breed and pass on their alleles. Both must be adapted to survive in the environment where they live.

ⓔ Student A makes good use of ideas about abiotic and biotic factors to explain that natural selection involves *all* the aspects of an organism's environment. This is unlike artificial selection where the selective agent (humans) chooses one feature (or a few features) and selection acts on one aspect of the organism. If the environment is stable, the same selection pressures operate from generation to generation. Student B does not refer to either environmental factors or selection. Both students could discuss what happens in a feature that shows continuous variation (see p. 15), which is possibly easier to explain than one showing discontinuous variation. The presence or absence of bands is involved with camouflage. Unbanded snails are found in uniform habitats such as grass. The absence of bands means that the snails blend in with their background and cannot be spotted by predators. Bands on the shell help to break up its shape. This is useful in habitats where there is variation in the vegetation, for example in hedgerows. Since populations of these snails live in places where there is a mixture of vegetation, which changes with the seasons, there is selection for maintaining both phenotypes and therefore both alleles. Student B fails to score.

ⓔ **Student B gains 5 marks out of 10 for Question 3.**

Question 4 **Cloning**

Micropropagation is a method of producing large numbers of plants using tissue culture. Cells taken from tissue culture can be genetically modified.

(a) Describe the process of micropropagation and outline how cells in tissue culture may be genetically modified.
In your answer you should make clear how the steps in each process are sequenced. (8 marks)

ⓔ This question covers two aspects of the specification for this unit and there is no introductory material in the form of text, a graph, a diagram or a photograph to help you. Before you start writing an answer, make a plan. Put some subheadings in the margin to make sure you cover both parts of the question.

(b) Explain what is meant by the term *non-reproductive cloning*. (2 marks)

Total: 10 marks

ⓔ Is this term in your glossary?

Student A

(a) An explant is taken from the growing tip of the plant. It is surface sterilised by placing into a solution of sodium hypochlorite. Aseptic conditions are necessary as fungi and bacteria could contaminate the growth medium. The tissue should be grown in a growth room where environmental conditions are controlled. Explants are transferred to sterile nutrient agar. The cells grow into a callus. Cytokinins at an optimum concentration are added to stimulate mitosis. The medium contains magnesium ions, nitrate ions and sucrose. Magnesium is needed to make chlorophyll, nitrate ions to make amino acids and sucrose is the substrate for respiration. Auxins are added later to promote root growth for stabilising the plant. The callus can be subdivided to form many groups of undifferentiated cells that are all genetically identical. Each cell has the ability to divide and differentiate into a complete plant with all the various specialised cells. This is useful if the cells have foreign DNA added, either by firing it into the cells (biolistics) or by using the T_i plasmid from *Agrobacterium* as a vector in genetically modifying plants, e.g. rice as in Golden Rice™. The gene(s) required is/are put into the plasmid which is taken up by the bacterium and then transferred to the chromosome of the cell in tissue culture.

Student B

(a) Sterile tissue is placed into a nutrient medium. The surface is sterilised to ensure that no disease or infection may harm the plant as that too would be cultured with undesirable effects. Plant growth regulators (e.g. cytokinin) are added into the culture medium to help form the callus tissue. The tissue is then subdivided to give many genetically identical pieces of callus. The contents of the medium all serve a purpose to create strong healthy plants. Sucrose is added as the main food supply for the growth of new plants. Auxin is then added to the growth medium to help growth by encouraging the cells to differentiate.

ⓔ Student B has missed the fact that there is a second part to the question that asks how cells in tissue culture are genetically modified. You have to know an outline of how rice is genetically modified — student A uses that information here. Student A could have used the term *totipotent* to describe cells that are capable of differentiating into any type of cell. Wherever possible, use appropriate technical terms. Student B does not gain the quality mark, as part of the answer is missing and the instruction in italics in the question makes it clear that both parts should be discussed. Student B gains 3 marks.

Student A

(b) This describes all the types of cloning that do not give rise to new individual organisms. This may involve cloning stem cells that are committed to becoming blood cells. This is used in research and for therapeutic cloning such as making β cells for people who have diabetes or genetically modified T cells for people with SCID. The mouse that grew a human ear is another example of non-reproductive cloning.

Student B

(b) Reproductive cloning is like plants that produce runners (strawberry) or suckers (elm). They grow little new plants that break off and become new individuals. Non-reproductive cloning is taking a fertilised egg, letting it divide and then splitting it to give two genetically identical embryos. These can grow again and the splitting process can be repeated. All the individuals are clones, but the cloning process happens after reproduction made the fertilised egg.

ⓔ Student B has wasted time not answering the question. The answer given is about reproductive cloning. Student A has given three examples, even though this is not required by the question. Once you have made sure that you have answered the question asked, it can be a good idea to give an example, especially in a question that asks for definitions. Student B fails to score.

ⓔ **Student B gains 3 marks out of 10 for Question 4. This question has no introductory information and no easy marks as there are in other questions in this section. You must learn something for each learning outcome and make sure you can write about it without any help from the exam paper.**

Question 5 **Ecosystems: fieldwork**

A group of students investigated the distribution and abundance of plants growing in a field and in adjoining woodland. They placed a tape measure between a point in the middle of the field and between the trees in the woodland.

Figure 4 shows a point quadrat.

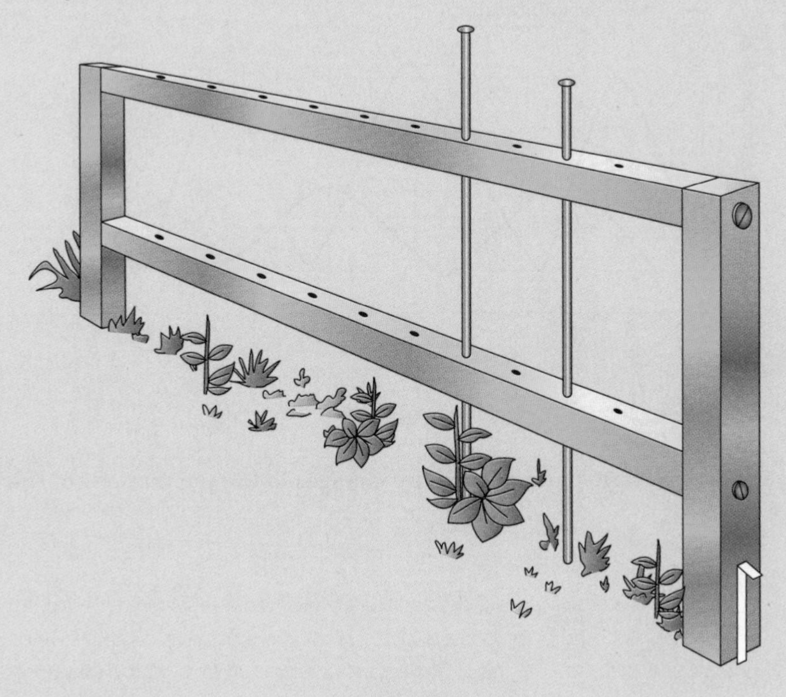

Figure 4

(a) Describe how the students should use a point quadrat to assess the distribution and abundance of plants in the field and the woodland. Explain how the data collected could be processed to show how the distribution and abundance of plant species changes from the field into the woodland. (4 marks)

e This is another question that could be answered as a set of numbered points written as it might be in a work sheet.

The cyclamen mite, *Phytonemus pallidus*, is a pest of strawberry crops in California. Infestations of these mites may be controlled by a predatory mite, *Typhlodromus reticulatus*.

In an investigation, the two species of mite were released on some strawberry plants. The numbers of both types of mite were recorded over a 12-month period. The results are shown in Figure 5.

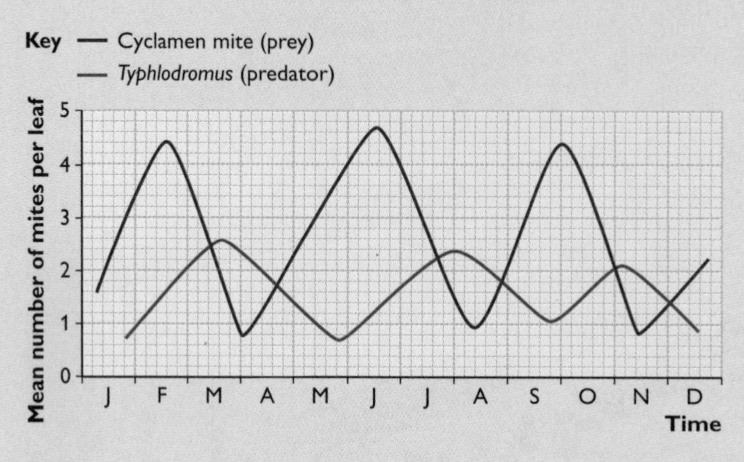

Figure 5

(b) Using the data in Figure 5, describe and explain the changes in the populations of the mites. (4 marks)

Total: 8 marks

e 'Using the data in...' is a clear instruction to translate information from the graph into prose in your answer. The question says 'changes' so you must give comparative data, e.g. the population increased from...at...to...at... Always include both axes in your data quotes. Use a ruler to make sure they are accurate.

Student A

(a) They record the first hit on each species made with each pin. The species touched by each pin are recorded. There are ten pins. The percentage cover is estimated by calculating the number of 'hits' as a percentage of the number of pins used. If four pins hit species A, then the percentage cover is $(4/10) \times 100 = 40\%$. You ignore multiple hits by each pin on the same species. The sampling is repeated at intervals along the tape measure. The results can be used to draw kite diagrams for each species along the transect. Bar charts could also be drawn for each sampling point.

Student B

(a) There are various sampling methods that a biologist could use to show the distribution of organisms across a habitat. You could first put a tape measure across the habitat from the field to the woodland and record the plant nearest the line every 50 cm. This would show how the vegetation changes across the habitat at a glance when you draw the results. You can then put the point quadrat at the same sampling points at right angles to the tape. You then put each pin through the frame and record the number of times each pin hits each type of plant. The results are expressed as the mean number of hits on each species per pin.

e Student A describes a simple method for using the point quadrat. The important point to remember is that with this method the *first* hit by the pin on each species is recorded and multiple hits are ignored. Student B has been taught the method of cover repetition. This involves recording how many times each pin hits each species on its way down. This method gives a good indication of the relative 'importance' of each species in a community in terms of its biomass or productivity. It is useful in field studies when researchers do not want to collect vegetation for weighing to find biomass and perhaps for burning to find the energy content. Student A's method is good to use in the field for a quick overview of the changes in the plant community. Cover repetition is good for more quantitative results. Always make sure you use the same method for recording the results of 'hitting' the plants with the pins. Fieldwork helps you to remember the various procedures and understand why they are carried out. There are several ways in which data can be recorded when using a point quadrat (see above). Student B scores 2 of the 4 marks. The answer has not been developed to show how the data can be used.

Student A

(b) When the cyclamen mites are put on the strawberries, they find plenty of food. They reproduce and their mean numbers increase from 1.6 per leaf in January to 4.4 in February as there are few predators and few limiting factors. Now there is more food for the predator mites so their numbers increase about a month later from 0.8 in Jan to 2.6 in March. As their population increases, the predators begin to eat more of the prey so the prey numbers decrease. Now there is less food for the predators and their death rate increases and they probably produce fewer offspring. The decrease in predators means that more prey survive to breed, so their numbers increase again. This cycle repeats. The maximum number of pest species (4–5 per leaf) is always higher than the maximum number of predators (2–3 per leaf).

Student B

(b) The numbers of cyclamen mites increased to 4.5 mites per leaf and then decreased over a 3-month period. The predator population was low to start with. The predators increase, but after the increase in the cyclamen mites. There is a maximum number of 2.5 mites per leaf and then the numbers decrease. This took 4 months. The same pattern is repeated twice more during the investigation.

ⓔ Both students describe the predator–prey model that occurs in a simple ecosystem with few species, but student B does not *explain* the changes. Here there are three species: strawberries, cyclamen mites and the predatory mite. This is similar to the much-quoted example of the snowshoe hare and the Canadian lynx. In most ecosystems, there are complex food webs — one predator species does not feed on only one prey species and each prey species has more than one predator. In reality, populations of primary consumers are controlled mainly by the availability of food, not by the presence of predators. Numbers of snowshoe hares (and many small rodent species) show a 10-year cyclical pattern dependent on the availability of food. Predator species are dependent on the populations of rodents that increase because there is more to eat, not because there are few predators. Student B gains 1 mark for describing the pattern in Figure 5.

ⓔ **Student B gains 3 marks out of 8 for Question 5.**

Question 6 **Plant responses**

Shoots are positively phototropic. Membrane proteins known as phototropins are thought to be involved as light receptors in the phototropic response.

The response to unidirectional light was investigated in thale cress, *Arabidopsis thaliana*, by using light of different wavelengths. The effectiveness of the different wavelengths in stimulating a phototropic response in seedlings of thale cress was determined and is shown in Figure 6. Phototropin was extracted from thale cress. The absorption of light of different wavelengths by a sample of phototropin was determined and is shown in Figure 7.

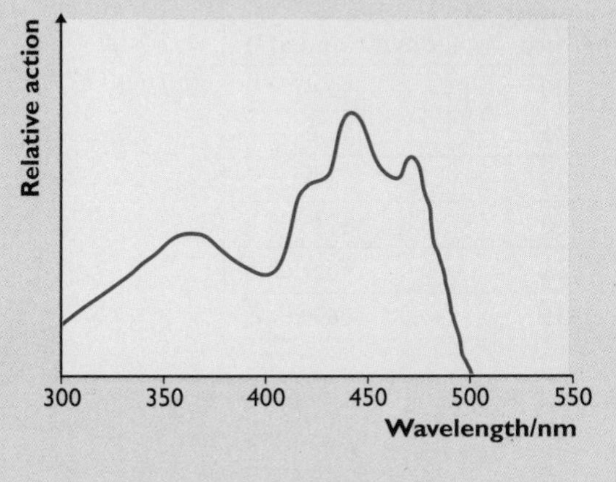

Figure 6

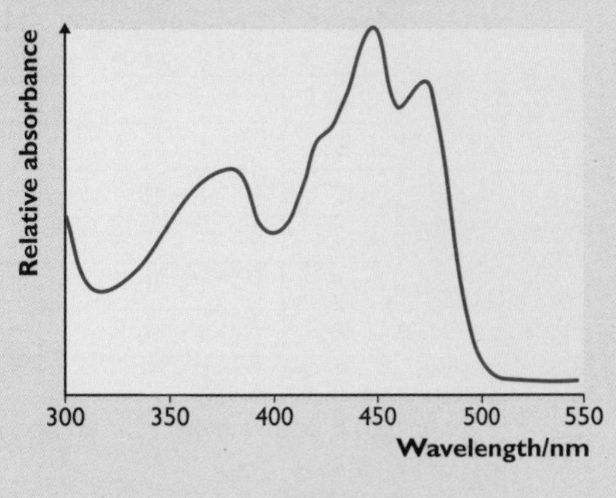

Figure 7

(a) (i) Describe the evidence shown in Figures 6 and 7 that supports the hypothesis that phototropin is the receptor in the phototropic response. (3 marks)

You must read the introductory information and study any graphs or tables before you read the questions. This ensures that you understand it all before you focus on what you are required to do in answering the questions. You should annotate the text and any graphs and tables to help you. This certainly will help answer this question.

(ii) Explain why it is necessary to be cautious when drawing conclusions based on evidence from Figures 6 and 7 alone. (2 marks)

This is a question on the *How Science Works* theme. Look very critically at the data provided and write down some ideas *before* you write out your answer.

The phototropic response involves auxins.

(b) Outline the way in which auxins are involved in the phototropic response of shoots. (3 marks)

This question tests your knowledge, but 'outline' means you are not expected to give much detail.

The gene, *Le*, codes for the enzyme 3β-hydroxylase, which catalyses the synthesis of the plant hormone gibberellin (GA_1) from its substrate GA_{20}. The recessive allele, le, codes for an enzyme that is not effective.

Dwarf pea plants with the genotype *lele* were grown from seed. Several weeks later, 60 plants all at the same developmental stage were divided into six equal groups, A to F.

The plants in each group were watered daily with 1 cm^3 of a solution of gibberellic acid (GA_3). After 5 weeks the stems of each plant were measured and means calculated for each treatment. The results are shown in Table 2.

Table 2

Group	Concentration of GA_3/g cm^{-3}	Mean length of stems/mm	Standard deviation (SD)
A	0	220	10.4
B	1×10^{-6}	176	16.8
C	1×10^{-5}	220	56.8
D	1×10^{-4}	360	137.2
E	1×10^{-3}	568	40.8
F	1×10^{-2}	536	66.4

e Read this information and the data very carefully.

(c) State three conditions that should be kept the same in this investigation. (3 marks)

e Make sure you do not give a condition that is given above, such as the volume of the solution added.

(d) Comment on the results of this investigation as shown in Table 2. (6 marks)

e 'Comment' invites you to *describe*, *explain*, *criticise*, *evaluate* and use the data to illustrate any points that you make.

Dwarf alleles have been incorporated into many crops, such as peas, wheat and barley.

(e) Suggest two advantages of growing dwarf varieties of crops. (2 marks)

Total: 19 marks

e These are always advantages to the farmer!

Student A

(a) (i) The absorption spectrum for phototropin and the action spectrum are almost identical. For example, the maximum absorption by phototropin occurs at a wavelength of 450 nm and the greatest phototropic response occurs at the same wavelength. There must be a receptor in the plant to detect the light that comes from one side. Since phototropin absorbs strongly in the same region of the spectrum that stimulates the biggest response, it suggests that phototropin is that receptor.

Student B

(a) (i) The receptors for the phototropic response are in the shoot tip. Figure 7 shows that phototropin absorbs light between 350 nm and 500 nm with a peak at 450 nm. Figure 6 shows that the most effective wavelength for stimulating the phototropic response is also 450 nm.

e Student B identifies the important evidence — the similarity between the absorption spectrum and the action spectrum — and refers to the peaks at 450 nm. However, the answer does not develop this idea in the way that student A's has. Student B gains 2 marks.

Student A

(a) (ii) There are two reasons to be cautious. There may be other compounds in the thale cress seedlings that could absorb light. The absorption spectrum is only for phototropin. The similarity in shape between the two graphs could just be a coincidence — phototropin could be absorbing light for another reason. Further evidence is needed.

Student B

(a) (ii) It is necessary to be cautious in drawing a conclusion because we are not told how often the experiment was carried out and whether there were any repeats. It could be that the results are anomalous.

e The question is worded to prompt answers that deal with the evidence in the two figures. Student B has given a general answer that would not gain any marks. Student A is correct, and further evidence is available. There are mutant alleles of the genes that code for the two types of phototropin. Seedlings of *A. thaliana* that are homozygous recessive for these genes do not show a phototropic response as the receptor is not functional.

Student A

(b) Auxins are produced in the shoot tip. They move down the stem. When light shines from one side they concentrate on the shaded side and stimulate elongation in the cells on that side. This occurs because auxins stimulate a membrane ATPase to pump H^+ out of the cells into the cell walls. This causes the cell wall to loosen so that the turgor pressure of water inside the cell causes the cell to increase in volume and stretch lengthwise.

Student B

(b) Shoots are positively phototropic and bend towards the light. Cells on the side opposite the light source, increase in size. This increase is stimulated by auxins.

ⓔ Student B states that cells increase in size, rather than increase in length. Only 1 mark could be awarded for the answer, but by not stating *length*, the student has lost that mark and fails to score.

Student A

(c) Light intensity, temperature, humidity

Student B

(c) Light, quantity of soil, water

ⓔ In questions of this type 'light' is not precise enough. Light duration (hours of light) and light intensity are acceptable answers. Quantity of soil is a good answer as the experiment is dependent on applying gibberellic acid to the soil. The volume of gibberellic acid solution applied is given in the question so that should not be quoted as an answer. The volume of water given to the plants each day would be a good answer, but 'water' alone is too imprecise. Student B should realise from the practical work for Unit F216 that more precise answers are required. Student B gains 1 mark.

Student A

(d) The dwarf plants cannot make GA_1 because the enzyme coded by the allele ***le*** does not function. Group A is the control — it is like the baseline. The others that were given GA_3 show greater length, apart from Group B. It could be that this is an anomalous result or that the difference between the means for A and B is not significant. GA_3 acts to replace GA_1 and stimulate growth in height. However, we do not know whether it completely replaces GA_1 because we do not have results for a group of tall plants (***LeLe***) for comparison. The standard deviations show that there was quite a range of stem lengths within each group, so it may not be possible to draw any conclusion about the effect of increasing the concentration of GA_3 on stem growth. They do show that it is possible to reverse the effect of the recessive allele.

Student B

(d) Plants that were watered with GA_3 have grown taller than those given water alone. Group A is the control that shows some growth although we are not told how much they have grown in the 5 weeks. One group watered with GA_3 has not grown as much as group A. The highest concentration ($1 \times 10^{-2}\,g\,cm^{-3}$) gave plants that grew less than the plants in Batch E so at high concentration GA_3 is inhibiting growth.

ⓔ The first sentence in student B's answer is incorrect. Student B is not sure how to respond to the command word *comment*. When you see this, try to think of different aspects to write about. You could describe the results, look for a trend (sketching a graph is useful if the data are presented in a table) and explain the results using your knowledge. The inclusion of standard deviations suggests that you could evaluate the results. You could also evaluate the procedure. The point about how much the plants have grown and not knowing the stem lengths at time 0 are valid. The plants will not all have been the same height. Note that the plants in Batch F still grew a lot. At that concentration, GA_3 is not inhibiting growth. Student B gains 1 mark.

Student A

(e) Crop plants have been bred so that they are dwarf. This means that there is less waste after harvesting stems and leaves and the plants will be able to put more energy into making grain.

Student B

(e) Dwarf plants put less energy into parts that we can't eat. Dwarf crops are less likely to fall over during storms because the stems are stronger.

e The collapse of crop plants as a result of storms or heavy rain is called lodging. Student B is correct that dwarf plants are less likely to show lodging. The first half of the answer however needs to be developed more to gain a mark. Student B scores 1 mark.

e **Student B gains 5 marks out of 9 for Question 6.**

Question 7 **Animal responses**

Figure 8 shows the arrangement of thick and thin filaments in a sarcomere from striated muscle as viewed (a) in longitudinal section (LS) and (b) in transverse section (TS) in an electron microscope.

(a)

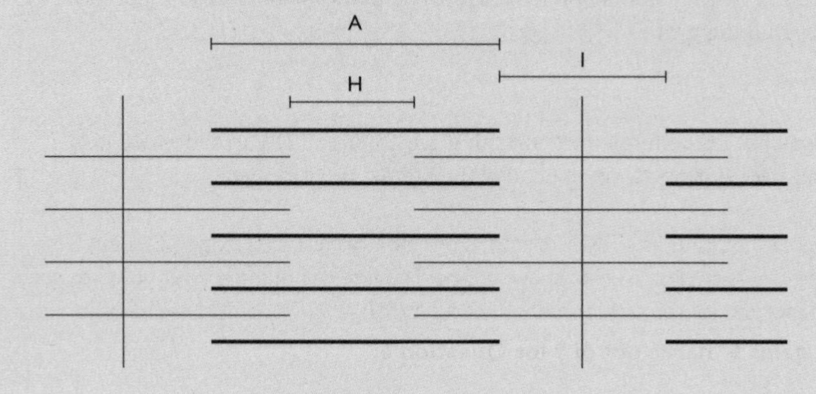

(b)

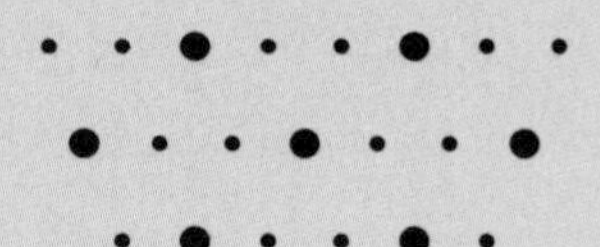

Figure 8

(a) Draw a line on Figure 8 (a) to show where the transverse section in Figure 8 (b) was taken. (1 mark)

This question would not have an answer line. Do not miss it.

(b) State what happens to the lengths of the following when the muscle contracts and shortens:

(i) A band

(ii) H zone

(iii) I band (3 marks)

Brief answers only are required here.

Aplysia californica **is a sea hare that has been studied to find out how the nervous system controls behaviour. Figure 9 shows a sea hare from the side and from above.**

When the siphon is stimulated by a jet of water the gill is withdrawn quickly into the mantle. If the siphon is stimulated ten times at 1-minute intervals the animal stops withdrawing its gill — a form of learned behaviour that lasts for a day.

(c) (i) Name the type of behaviour described above. (1 mark)

(ii) Explain the advantage of this type of behaviour to *Aplysia*. (2 marks)

Think in terms of energy.

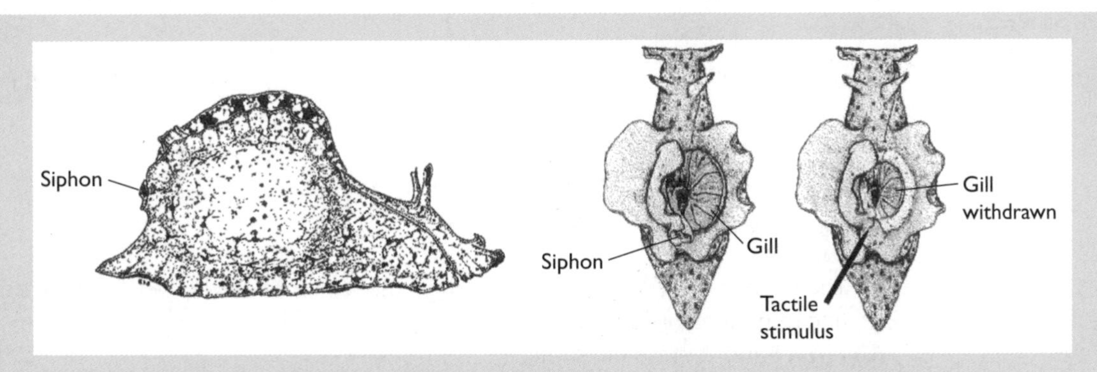

Figure 9

Figure 10 shows the neural pathway involved in this behaviour and the membrane potential in the sensory neurone and the post-synaptic membrane potential at X on the motor neurone when the siphon is stimulated.

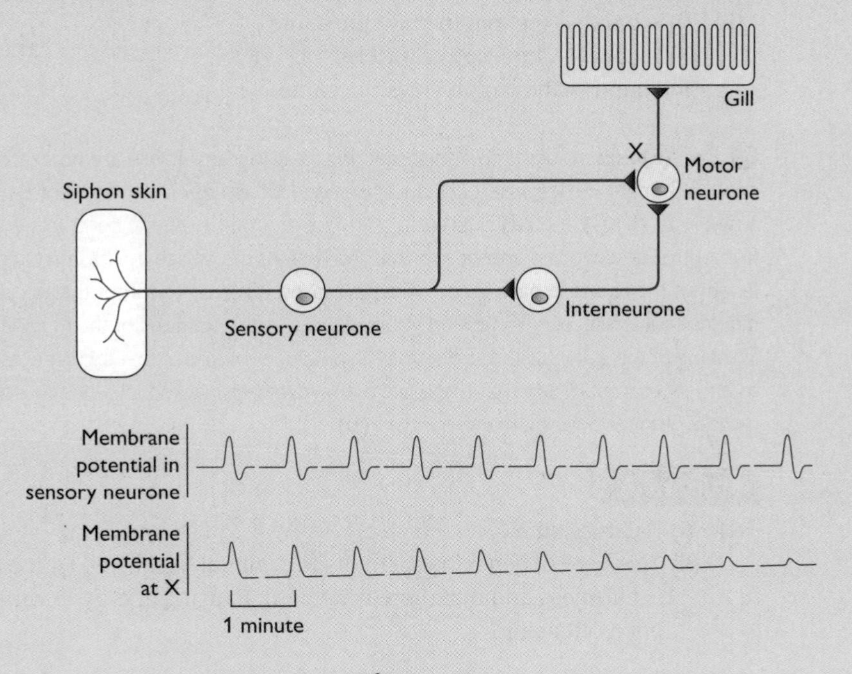

Figure 10

(d) Suggest how the neural pathway shown in Figure 10 is responsible for the change in gill withdrawal behaviour on repeated stimulation of the siphon. (3 marks)

You should be able to work out the answer by careful analysis of the information provided.

(e) In dangerous situations a cat will puff up its tail, arch its back, and assume a sideways position indicating a 'fight-or-flight' response. Explain how this response is coordinated in mammals, such as a cat. (5 marks)

Total: 15 marks

ⓔ 'Coordinated' should make you think of the nervous and hormonal systems. Think about the coordination, not the responses of the cat, when answering.

Student A

(a) The student drew a vertical line through the region of overlap between the thick and thin filaments.

(b) **(i)** A band — stays the same

(ii) H zone — gets smaller

(iii) I band — gets smaller

Student B

(a) The student did not draw anything on Figure 8.

(b) **(i)** A band — the length stays the same

(ii) H zone — the length shortens

(iii) I band — the length stays the same

ⓔ Part **(a)** asks students to add something to a diagram. There are no dotted lines for a written answer. Student B has missed this question and lost an easy mark. Student B has the correct answers to **(b) (i)** and **(ii)** but has forgotten that when a muscle contracts the thick and thin filaments slide over one another so that the regions of overlap increase and correspondingly the length of the I band in each sarcomere decreases. It is easy to forget that myofibrils are three-dimensional structures. To remind yourself, draw cross-sections of the H zone and the I band. Thinking of the Z line as the Z disc may also help. Remember that sarcomeres are arranged end to end in a myofibril and that there are many myofibrils packed into one muscle fibre of striated muscle. Student B gets 2 marks for part **(b)**.

Student A

(c) **(i)** Habituation

(ii) This type of behaviour benefits the animal because it learns that the stimulus is harmless and that there is no point in using energy to withdraw the gill and put it out again.

Student B

(c) **(i)** Classical conditioning

(ii) The animal is learning that there is nothing to be feared by the stimuli on the siphon. If it was 'nervey' and retracted parts of its body every time it was stimulated even slightly then it would never get enough food, find shelter or meet a mate.

e Student A recognises that this is habituation. Student B confuses this behaviour with conditioning. Only one stimulus is mentioned in the question, so this cannot be classical conditioning. In their different ways both students have expressed the right ideas to part **(ii)** and student B gains 2 marks.

Student A

(d) As the siphon is stimulated, the sensory neurone transmits impulses to the motor neurone and an interneurone. The post-synaptic potential in the motor neurone decreases over this time. This could be because the interneurone is an inhibitory neurone and causes a hyperpolarisation in the cell body of the motor neurone. This makes it more difficult for the sensory neurone to stimulate the motor neurone.

Student B

(d) There are several ways in which this might happen, but it seems likely that there is a change in the synapse between the sensory neurone and the motor neurone. Perhaps with repeated stimulation the neurotransmitters at the end of the sensory neurone are used up so no impulses can be sent to the motor neurone.

e In this case, student B is closer to what happens in this neural pathway in habituation. However, the command word in the question is 'Suggest' and this means that any likely mechanism will be accepted by the examiners, so both students gain the full 3 marks. This question relies on information from Unit F214 about transmission along neurones and across synapses. In this examination, anticipate questions that rely on information from the other A2 unit as well as from the two AS units.

Student A

(e) During a stressful situation the sympathetic nervous system (SNS) stimulates the adrenal medulla to secrete adrenaline. SNS neurones innervate many organs in the body to prepare for fighting or running away. The release of noradrenaline by SNS neurones and the stimulation by adrenaline:

- increases heart rate
- increases the stroke volume
- diverts blood from the gut to muscles by stimulating vasoconstriction in the gut and vasodilation in the muscles
- increases the width of the airways in the lungs

Adrenaline activates an enzyme cascade in the liver to break down glycogen to glucose. The glucose diffuses into the blood and this raises the glucose level. There is more energy available to the muscles.

Student B

(e) The cat's fight-or-flight response is coordinated by the nervous system and adrenaline from the adrenal glands. Nerves stimulate the SA node in the heart so it beats faster. Adrenaline stimulates the liver to produce glucose. Motor nerves stimulate the hairs in the skin to be raised by contraction of muscles. Sensory awareness is heightened — for example the pupils become wider.

ⓔ Student A has used bullet points, which is a good way to list the ways in which adrenaline and the sympathetic nervous system (SNS) increase the readiness to fight or run away. Student B's answer is more vague. There is nothing about the SNS and the role of adrenaline in increasing the blood glucose concentration lacks detail. Student B scores 3 marks for the effects on the heart, skin and pupils. He/she has missed the opportunity to refer to synoptic material on adrenaline and the control of the heart by the autonomic nervous system from Unit F214.

ⓔ **Student B gains 10 marks out of 15 for Question 7.**

Overall, student B gains 38 marks. This may not be enough for an E grade. You can see that student B has lost marks for a number of different reasons:

- **Some answers are not developed fully, e.g. Q1(a) and (b), Q5(a), Q6(a)(i) and (e).**
- **Terms from the specification are used incorrectly, e.g. Q7(c) and (e).**
- **Terms have not been explained, e.g. Q4(b).**
- **Information has been described, rather than explained, e.g. Q2(e).**
- **Not understanding what is required for an answer, e.g. Q1(a), Q2(d)(ii), Q5(a).**
- **Instructions have not been followed carefully, e.g. Q2(a) and Q6(d) where the command word comment was not understood.**
- **Data provided have not been used fully, e.g. Q1(c)(ii).**
- **Answers are not precise enough, e.g. Q1(c)(i).**
- **The 'full story' has not been given, e.g. Q6(b).**
- **In questions that address *How Science Works*, not answering the question to explain how data support or refute statements, e.g. Q3(d), and not being precise enough with naming variables in Q.6(c). Remember that 10% of the marks are allocated to this theme.**
- **Questions are missed, e.g. Q1(c)(ii), Q4(a) and Q7(a).**
- **Irrelevant material is included, e.g. Q3(b).**
- **Facts have not been learnt, e.g. Q4(b).**
- **Data on the exam paper have been read incorrectly, e.g. Q3(a).**
- **A previous answer has been repeated, e.g. Q2(d)(ii).**
- **All the steps have not been given, e.g. in the genetic diagram in Q2(b).**
- **Synoptic material has not been used to best effect, e.g. Q1(b) and Q7(e).**

Knowledge check answers

1

Amino acid	Coding polynucleotide	Template polynucleotide
Histidine	CAT, CAC	GTA, GTG
Phenylalanine	TTT, TTC	AAA, AAG

You may find other terms used for the polynucleotides (or strands) of DNA:
- coding — sense
- template — antisense, reference

2 A two-base code would only code for (4^2) 16 amino acids.

3 met–leu–ala–ile–ala

4 Reading errors may give rise to a different sequence of amino acids — especially if they are in the first base of a codon.

5 DNA has two antiparallel polynucleotides. 3' and 5' indicate the different ends of a polynucleotide. They refer to the carbon atoms on the sugar, deoxyribose.

6 The deletion of the fifth base pair changes the sequence of codons after the start codon (AUG). The sequence of amino acids after met is completely different; this is likely to code for a non-functional polypeptide.

7 The mRNA is AUGAUACGGCUUACGUUAG… which codes for: met–ile–arg–leu–thr–leu–… . The sequence of amino acids is very different and there is no stop codon so the polypeptide will be longer than the original.

8 The wrong amino acid would be inserted into the primary sequence of the polypeptide. This type of mutation is so catastrophic that they are usually lethal early in development. Mutations in other regions of genes for tRNA occur in mitochondria, which contain DNA that codes for their own tRNAs. These are related to some of the 150 known human disorders associated with mitochondria (see p. 49).

9 In both (a) and (b), glucose is present in the medium so CRP does not bind to the DNA as shown in Figure 5(a). This means that there is no transcription of the structural genes. If there is no lactose then the repressor protein binds to the operator site, blocking transcription.

10 If there is no glucose, then cAMP increases in concentration and binds to CRP. CRP–cAMP binds to the CRP site making it possible for RNA polymerase to bind to the promoter site. If there is no lactose, the repressor protein still binds to the operator and blocks transcription.

11 They can be considered to be codominant at the biochemical level since both alleles code for polypeptides that function in red blood cells. As it is often difficult to tell whether someone has sickle-cell trait ($\mathbf{Hb^AHb^S}$) or not ($\mathbf{Hb^AHb^A}$) without doing a biochemical or genetic test then $\mathbf{Hb^A}$ may be said to be dominant. However, people with sickle-cell trait may experience symptoms. Heterozygotes have a resistance to malaria that is not shared with people who are $\mathbf{Hb^AHb^A}$.

12
Group A $\mathbf{I^AI^A}$, $\mathbf{I^AI^O}$
Group B $\mathbf{I^BI^B}$, $\mathbf{I^BI^O}$
Group AB $\mathbf{I^AI^B}$
Group O $\mathbf{I^OI^O}$

13 Blood groups AB and Rhesus positive. The Rhesus locus is on chromosome 1 and controls another blood group antigen (D). Rh+ have antigen D on their red cells, Rh– do not. The allele **R** (Rh+) is dominant to **r** (Rh–). The inheritance of both blood groups would make a good question about independent assortment.

14 Chromatid: one of the two 'strands' that make up a chromosome following replication. Each chromatid of a double-stranded chromosome is made of a molecule of DNA and molecules of proteins.
Sister chromatids have identical (or near identical) DNA and are joined at the centromere. (Gene mutation may have occurred during replication in which case the sister chromatids will be slightly different.) At the beginning of anaphase in mitosis and anaphase II in meiosis, sister chromatids break apart and become daughter or single-stranded chromosomes. Non-sister chromatids are the chromatids of the two chromosomes in a homologous pair; they are not joined at a centromere. Chiasmata form between non-sister chromatids.

15 Mutation; other causes of variation are covered in the section on meiosis and variation.

16 F_1 red-eyed males ($\mathbf{X^RY}$) and red-eyed females ($\mathbf{X^RX^r}$); F_2 red-eyed males, white-eyed males and red-eyed females in a ratio of 1:1:2. Of the females, 50% are homozygous dominant and 50% are heterozygous.

17 Three boys inherited the grandfather's X chromosome through their mother, who is a carrier. The fourth has inherited an X chromosome produced by crossing over, which occurred in the boy's mother. Look out for questions on sex linkage that involve animals, such as butterflies, moths and birds, in which the females are XY and males are XX. Look out also for genes that are on the Y chromosome, such as the sex-determining gene (*Sry*) in humans.

18 $\chi^2 = 5.38$. With three degrees of freedom $p > 0.05$ so the difference between observed and expected results is not significant.

19 Variation is reduced.

20 Genetic variation between individuals of the same species is related to the different *alleles* that they have. All individuals of the same species have the same genes.

21 Einkorn and wild grass (once thought to be *Aegilops speltoides*) 7; emmer wheat 14; spelt and modern wheat 21.

22 (a) p (**F**) = 0.98; q (**f**) = 0.02 (b) = 3.92, which should be rounded to 4% of the population (1 in 25).

23 A ($p^2 + 2pr$) = 0.45; B ($q^2 + 2qr$) = 0.13; AB ($2pq$) = 0.06; O (r^2) = 0.36

24/25 The figures you calculate should give a straight-line graph.

26 They involve culturing microorganisms in artificial conditions.

27 256

28 Four (dATP, dTTP, dCTP, dGTP)

29 Annealing is the formation of hydrogen bonds between complementary base pairs. Ligation is the joining of DNA fragments by formation of phosphodiester bonds.

30 Grazing food chain: grass → rabbit → fox
Decomposer food chain: dead grass → fungi and bacteria

31 Detritivores 'shred' the food they eat, absorb some of it and egest the rest with a large surface area for decomposers to act on. Decomposers break down organic material, recycling carbon as carbon dioxide and nitrogen in organic compounds (proteins) as ammonia.

32

Microorganism	Role in nitrogen cycle
Decomposers (bacteria and fungi)	(See answer to Knowledge check 31)
Nitrifying bacteria (*Nitrosomonas* and *Nitrobacter*)	Convert ammonia to nitrate ions (see page 56)
Nitrogen-fixing bacteria (*Rhizobium*)	Convert dinitrogen (N_2) into ammonia
Denitrifying bacteria	Convert nitrate ions to dinitrogen

33 There is a constant supply; replanting replaces the trees that are felled.

34 Escape reflex

35 Innate behaviour is inborn; the same response is always given to the same stimulus. Learned behaviour is acquired during an animal's lifetime.

36 People are diploid, with one pair of homologous chromosomes (in this case No. 11).

A

abiotic factors 51, 52, 54, 88
ABO blood group 16
abscission 62
allele frequency
  calculation of 33–34
  changes in 28–29
  genetic drift 30
  and selection 31–32
alleles 20–21
amino acid activation 9
amino acids, triplet code 7
anaphase I 18, 19
anaphase II 18, 19
animals
  behaviour 69–71
  cloning of 37–38
  genetic engineering 50
  responses 62–69, 100–04
antagonistic muscle action 66
antibiotic resistance genes 42, 47, 49
anticodons, tRNA 9
apical dominance 61
apoptosis 13–14
artificial ecosystems 53–54
artificial selection 31–32
autonomic nervous system 64–65
auxins 60–61, 62

B

β-galactosidase 12, 74
bacteria
  genetic engineering 45–48, 49
  growth of, culture 39
  nitrogen cycle 55, 56
base pair changes 10–11, 71
bases, DNA 7
batch cultures 40
biospecies 30
biotechnology 38–40
biotic factors 52, 88
blood group systems 16, 21
brain structure and function 65–66
bread wheat, evolution of 31–32

C

callus culture 36, 89, 90
cAMP (cyclic AMP) 12, 13, 14
cardiac muscle 64
carriers, sex linkage 24, 25
carrying capacity 57
cell death 13–14
cerebellum 65, 66
cerebrum 65
chain termination method 44, 83
chi-squared test 22–24
cladistics 30
classical conditioning 70
classification of species 30
clear felling 57–58
cloning 35–38, 89–90
CNS (central nervous system) 63, 64, 65, 66
codominance 21
codons, mRNA 7–8
communities 52, 54
competition, species 57
conditioning 70
conservation 57
constitutive enzymes 11–12
consumers 51, 52–54
continuous cultures 40
continuous variation 15, 16
coppicing 58
critical value 23
crops
  energy efficiency 53–54
  genetic modification 47–48, 49
  use of plant hormones 62
crossing over 25–26
CRP (cyclic AMP receptor protein) 13
cyclic AMP (cAMP) 12, 13, 14
cytokinins 61, 62

D

Darwin, Charles 28, 58
decomposer food chain 52
decomposers 52
decomposition 54–56
degenerate genetic code 7
DELLA proteins 62
denitrification 55, 56
detritivores 52, 54
dihybrid crosses 21, 22
directional selection 29
discontinuous variation 15, 16
DNA probes 43, 46
DNA replication 7, 10, 41, 45
DNA triplet codes 7
dopamine receptor D4 (*DRD4*) 70–71
downstream processing 40
Drosophila melanogaster 72
  crossing over in females 25–26
  embryo development 13–14
  eye colour 24–25
  wing length gene 20–21
dwarf plants, gibberellins 32, 61–62, 98–99

E

ecological efficiency 53
ecosystems 51–59
electrophoresis 42, 85, 86

energy flow, ecosystems 51–54
environmental resistance 57
environment-gene interaction 16–17
enzymes 9, 11–12, 13, 14
  biotechnology processes 40
  and epistasis 26–27
  genetic engineering 45–48
  gibberellin pathway 61–62
  restriction 41, 45, 46
  reverse transcriptase 42, 46
epistasis 26–28
Escherichia coli 12, 13, 74, 75
ethical issues 50
eye colour, sex linkage 24–25

F

fermentation 40
fieldwork 91–94
'fight or flight' response 68–69
'fixed' nitrogen 55
flower pigments, epistasis 27–28
fluorescent dye, DNA 43, 45, 83
food chains and webs 51, 52
food production
  energy efficiency 53–54
  microorganisms 38–39
forearm, movement of 66
founder effect 30
frameshift mutations 11
fruit fly *see Drosophila melanogaster*

G

Galápagos Islands
  conservation case study 58–59
  directional selection 29
gel electrophoresis 42
gene pool 33
genes 7–8
  mutations 10–12
  and operons 11–13
  protein synthesis 8–10
  role in development 13–14
gene sequencing 41, 44–45
  questions & answers 83–87
gene therapy 48–49
genetic code 7–8
genetic diagrams 20, 21
genetic disorders 34, 48–49
genetic drift 30
genetic engineering 42, 45–50
genetic variation 15–28
genome 41
genotype
  codominance 21
  contributing to phenotype 16–17
  frequency of 33, 34
  and inheritance 20–21
  and linkage 26
  questions & answers 78, 79–80
  and sex linkage 24–25
genotypic variation 15
germ-line gene therapy 48, 49, 50
gibberellins 32, 61–62, 98–99
Golden Rice™ 47–48
grazing food chains 51
gross primary productivity 52
ground finch, directional selection 29

H

habitats 52
habituation 69, 102–03
haploid 19
Hardy–Weinberg principle 33–34, 84, 87
height of plants 61–62, 98–99
hemizygosity 24
herbicide-resistance genes 49
herbivores, nitrogen cycle 55–56
homeobox genes 13, 14, 45
hormones, plant 60, 61–62
horticulture, plant cloning 37
Hox genes 13–14
hybridisation 31, 32
hypothalamus 65, 66

I

imprinting 70
inbreeding 35
independent assortment 17, 22, 26
indole acetic acid (IAA) 60, 61
inducible enzymes 12
inheritance 20–21
innate behaviour 69
insight learning 70
insulin 8, 10
interspecific competition 57
intraspecific competition 57
intraspecific variation 15

K

kineses 69

L

lac operon 12–13
lactose 12–13, 74
latent learning 70
learned behaviour 69–70
limiting factors, population growth 39, 56–57
linkage 25–26
liposomes 41
livestock, productivity 53–54

M

mathematical skills
 chi-squared test 22–24
 Hardy–Weinberg principle 33–34
medulla oblongata 65, 66
meiosis 17–22
meristem culture 36
metaphase I 18, 19, 21, 22
metaphase II 19
microorganisms 16
 in biotechnology 38–40
 genetic engineering 49
 role in nitrogen cycle 55–56
milk yield 16–17, 32, 79, 82
mockingbirds 59
motor neurones 64, 67, 68, 70, 101, 103
mRNA (messenger RNA) 8, 9, 11, 12, 49
muscle contraction 67–68
 questions & answers 100, 102
muscle types 63–64
muscular movement, control of 66
mutagens 10
mutations 10–11, 71
mycoprotein production 38, 40
myofibrils, muscle contraction 67–68

N

natural selection 28, 32
negative control 13
nervous system
 autonomic 64–65
 organisation of 63
net primary productivity 52–53
neurotransmitters 70–71
neutral mutations 11
niches 52
nitrifying bacteria 56
nitrogen cycle 55–56
nitrogen-fixing bacteria 56
non-reproductive cloning 35–36
non-sister chromatids 17, 19, 25–26
nutrient cycling 54–56

O

operant conditioning 70
operator region 12
operons 12–13, 74, 75
organ transplantation 48

P

parasympathetic nervous system 63, 64, 65
Pavlov's dogs 70
PCR (polymerase chain reaction) 41, 43–44
performance testing 32
phenotype
 and codominance 21
 and inheritance 20–21
 and linkage 25–26
 questions & answers 79–88
 and sex linkage 24–25
phenotypic ratios, epistasis 28
phenotypic variation 15
photosynthesis and energy flow 52–53
phototropic response of shoots 60–61
phototropins 60, 95
phylogenetic species 30
pigs, organ donors 48
plant hormones 60, 61–62
plants
 cloning of 36–37
 energy flow from 52–53
 genetic engineering 49
 nitrogen cycle 55
 responses 60–62, 95–99
plasmids 42, 46–48
point quadrats 91, 93
polymerase chain reaction (PCR) 41, 43–44
polypeptides 7
 mutation of 10–11
 translation 9–10
polyploidy 31
populations 52
 isolation of 30
 and sustainability 56–57
positive control 13
predation, control of 54
predators and prey 57, 92–94
preservation 57
primary consumers 52–54
primary metabolites 38
primary succession 54
primates, social behaviour 70
probability table, chi-test 24, 79
producers 51, 53–54
productivity 52–54
progeny testing 32
promoter region 12, 13
prophase I 18–19
 crossing over during 26
prophase II 18, 19
protein synthesis 8–10
Punnett square 20, 27, 33, 80, 81

Q

quadrats 91, 93
quaternary structure, proteins 7, 10
Quorn™ 38, 40

R

rDNA (recombinant DNA) 47
reflex arc 66
reflexes 69, 70
regulator gene 13
reproductive cloning 35
restriction endonucleases 41, 45
restriction enzymes 45–46, 85–86
reverse transcription 42, 46
ribosomes 8, 9–10
risks of genetic engineering 49–50
RNA
  mRNA 8, 9, 11, 12, 49
  tRNA 9–10
RNA interference (RNAi) 49
RNA polymerase 12, 13

S

sampling methods 92–93
saprotrophic nutrition 54
sarcomeres, muscle contraction 67–68, 100, 102
SCID, treatment of 48–49
secondary consumers 53
secondary metabolites 38, 40
segregation of allelic pairs 20
selection 28–34, 88
selective breeding 32
selective felling 58
semi-conservative replication 7
senescence 62
sequencing a genome 44–45, 83–84
sere 54
sex linkage 24–25
sickle-cell anaemia 16
sigmoid growth 39, 40
significance, statistical 23–24
'silent' mutation 10, 11
sister chromatids 17, 19
smooth muscle 63, 64
somatic gene therapy 48
somatic motor system 64, 65
somatic nervous system 63
specialisation of genes 13–14
species definitions 30
Sr2 gene 32
stabilisng selection 28–29
'sticky end' 45–46
stop codons 7, 10, 11
striated muscle 64, 67, 100
strip-felling 58
structural genes 12
stutter mutations 11
substitution mutations 10
succession 54
superovulation 37
suspension culture 36
sustainability 56–57
sympathetic innervation 65, 68–69
sympathetic nervous system 63, 64, 65

T

taxes 69
telophase I 18, 19, 22
telophase II 18, 19, 22
template polynucleotides 7, 8
templates 7
tertiary consumers 51, 53
test crosses 22–23, 26
therapeutic cloning 38
timber production 57–58
tissue culture 36, 37, 89
transcription 8, 12
  inhibition of 13
  reverse 42, 46
transcription factors 13, 45, 62
transformed bacteria 47
transgenic bacteria 47
translation 9–10
tree felling 57–58
triplet codes, DNA 7–8
tRNA (transfer RNA) 9–10
trophic levels 51, 52–54
tropisms 60

U

unidirectional light, effect on IAA 61
unlinked genes, inheritance of 20–21

V

variation 15–28
vectors, gene cloning 41, 46–47
vegetative propagation 36
vegetative reproduction 36
visceral nervous system 63

W

wheat, artificial selection 31–32
wing length, inheritance of 20–21

X

xenotransplantation 48